RADIO PRODUCTION WORKTEXT
Third Edition

RADIO PRODUCTION WORKTEXT

Studio and Equipment

Third Edition

DAVID E. REESE
LYNNE S. GROSS

Focal Press
Boston Oxford Johannesburg Melbourne New Delhi Singapore

Focal Press is an imprint of Butterworth-Heinemann.

Copyright © 1998 by Butterworth-Heinemann

 A member of the Reed Elsevier group

Library of Congress Cataloging-in-Publication Data
Reese, David E.
   Radio production worktext : studio and equipment / David E. Reese,
Lynne S. Gross. — 3rd ed.
   p. cm.
   Includes bibliographical references and index.
   ISBN 0-240-80283-7 (pbk. : alk. paper)
   1. Radio stations—Equipment and supplies. 2. Radio—Production
and direction. 3. Sound—Recording and reproducing—Equipment
and supplies. I. Gross, Lynne S. II. Title.
TK6557.5.R44 1997
621.389'3—dc21   97-23429
                                                 CIP

British Library Cataloguing-in-Publication Data
A catalogue record for this book is available from the British Library.

The publisher offers special discounts on bulk orders of this book.
For information, please contact:
Manager of Special Sales
Butterworth-Heinemann
225 Wildwood Avenue
Woburn, MA 01801-2041
Tel: 617-928-2500
Fax: 617-928-2620

For information on all Focal Press publications available, contact our World Wide Web home page at: http://www.bh.com/focalpress

10 9 8 7 6 5 4 3 2 1

Printed in the United States of America

For "students" of broadcasting, whether in a formal class or studying on your own, we appreciate your desire to learn more about the equipment and techniques that make it possible to produce good radio production.

D.E.R.
L.S.G.

Contents

5 AUDIO TAPE RECORDERS

6 AUDIO TAPE EDITING

7 MONITOR SPEAKERS

8 CONNECTORS, CABLES, AND ACCESSORIES

11 MULTITRACK PRODUCTION TECHNIQUES

12 PRODUCTION TIPS, TRICKS, AND TECHNIQUES

APPENDIX A

APPENDIX B

Additional Production Projects

Preface

It has been almost ten years since *Radio Production Worktext* was initially written. This third edition follows the concept of the previous editions by offering a solid background of the basic techniques and mechanics of modern radio production and will be valuable for those who wish to become engaged in radio production or those who are already producing for radio but wish to know more about equipment and production techniques. Some production texts seem to offer burdensome amounts of information and explanations that don't have practical application in the day-to-day world of radio, while other texts wander into aspects of radio broadcasting that have only vague implications toward radio production. This text sticks to the subject at hand, and while radio production can seem complex to the beginner, *Radio Production Worktext* simplifies the understanding of the radio production process.

Radio is a very competitive field, and those with the most knowledge (and talent) stand the best chance of succeeding. Reading this text will get you started. However, as the title implies, this isn't a book just to read, but rather a text that will become an integral part of your study and practice of radio production. The worktext approach is an admission that this subject is really only taught by a combination of theory and discussion *and* practical, hands-on use of broadcast equipment. Each chapter of the text is divided into a section of **Information**, a section of **Self-Study Questions** and **Answers**, and a section of **Projects**.

The Information sections are further divided into modularized units so that all the primary concepts can be easily learned. These concepts are illustrated with drawings and photographs that help the student grasp the specific ideas being presented.

The Self-Study Questions include multiple choice, true/false, and matching questions covering the material presented in the chapter. Answers, as well as suggested procedures for those who don't answer correctly, guide the student through this section. These questions are intended to be instructional so that by the time the student has read the Information section and answered the questions correctly, he or she will feel confident about knowing the information.

The chapters' Projects sections contain practical projects that should give readers a degree of expertise in equipment manipulation and hopefully lead them on to further, creative radio production work. It is assumed that the reader has completed both the Information section and Self-Study Questions before beginning the various projects. The importance of practical production experience should not be overlooked, and these projects should be viewed as a starting point for the developing radio production person. Additional hands-on work in the production studio is never wasted time and is highly recommended.

The book is organized primarily by equipment, but it doesn't deal strictly with a nuts-and-bolts approach. Techniques as well as mechanics are included within each chapter. In fact, the book isn't intended to be technical. It's written in an easy-to-read style that should enable people without technical training to understand the nature and makeup of the radio process.

The third edition of this book continues to refine its organization and content, especially in light of radio production's transition from analog to digital technology and techniques. As a result, we have expanded the sections on digital equipment and procedures throughout the book. For example, while we still look at "cut and splice" editing in a chapter entitled "Audio Tape Editing," there's a new section on digital audio editing. Also, because of the continuing acceptance and development of digital technologies, the chapter on "The Digital Production Studio" has been updated and enlarged.

At the suggestion of some of our readers, we've added a chapter that deals with "Multitrack Production Techniques." This is an expansion of several sections in last edition's "Production Tips, Tricks, and Techniques" chapter, recognizing the importance of multitrack work that's readily accomplished with many digital technologies.

The third edition of *Radio Production Worktext* now includes an audio compact disc that illustrates many of the important concepts discussed in the text. For example, when the terms "cross-fade" and "segue" are explained in Chapter Two, not only is there a definition of the terms and a diagram showing the concept, but there's also an audio clip that lets you readily hear the difference between these two sound transitions. Those sections of the text that do come with an "audio clip" are noted throughout the text with a small "compact disc" icon.

Throughout the book we have freshened and updated both copy and artwork, but much of the basic information remains the same. As with the second edition, the first chapter introduces the reader to the layout and design of the radio production studio and describes briefly the equipment that's detailed in succeeding chapters. The initial chapter also discusses some basic concepts regarding "sound" that will help the production person to understand the "raw material" being worked with.

Chapter Two introduces the audio console, the heart of all production work. New material on "virtual" consoles brings this chapter into the digital age. A firm understanding of this chapter is important because most other broadcast equipment operates through the audio console. Having mastered that, students should be ready to work with the sound sources that can serve as inputs or outputs for the console. One of these, microphones, is covered in

Chapter Three. Chapter Four deals with CD players (a basic digital technology) and turntables.

Chapter Five looks at the audio tape recorders found in the production studio—the traditional reel-to-reel, cassette, and cartridge recorders, and the newer digital tape recorders. The size of this chapter is attributed to radio production's heavy reliance on audio tape recording and playback. Chapter Six deals with audio tape and audio tape editing—both analog and digital.

Often overlooked equipment is included in Chapters Seven and Eight. Monitor speakers and headphones are discussed in Chapter Seven, and Chapter Eight surveys connectors, cables, and accessories that complement the major pieces of equipment in the production studio. Chapter Nine provides a survey of the most popular signal processing equipment employed in radio production work. Once again you will learn that digital equipment can often replicate many of the signal processing effects.

The concept of a totally digital production studio—in fact, a totally digital radio station—is introduced in Chapter Ten. Most current production studios house at least some digital equipment, such as a compact disc player, a minidisc recorder, or perhaps a digital editing system. The future importance of digital technology in radio production should become apparent as you learn about digital cart recorders, computer audio editors, and digital audio workstations in this chapter.

Chapter Eleven, as mentioned previously, is a new chapter on multitrack production techniques. While some analog-based production studios have multitrack capability, almost all digital-based studios do, and it's important for the production person to learn the basic techniques of multitrack recording and the creative possibilities they open up.

The last chapter explains some basic production-room techniques plus some selected production tricks that should get you thinking further about creative production and how you can apply all the ideas you've learned as you've worked through this worktext.

Appendix A provides a brief explanation of various production situations—such as news, commercials, and sports play-by-play—in which you can employ some of the production skills mentioned in this text. Appendix B provides several additional projects that can be appropriately tackled after completing the text.

Throughout the book, key terms are listed in **boldface** when first introduced. These terms are explained in the text and definitions are also included in the glossary at the end of the book. A selected list of additional reading is provided for students who wish to continue their study in this area.

The authors are indebted to many people for their help and encouragement as the third edition of *Radio Production Worktext* was being put together. Our colleagues within the Communications Departments at John Carroll University and California State University, Fullerton, offered generous support for this project's realization. Many additions to the third edition of this text are a result of the helpful comments that came from instructors using the first two editions. A special "thanks" goes to our reviewers for this edition—Carla Gesell from University of Tennessee at Martin and Dave Evans from WKDQ, Kentucky. Their comments and suggestions were very helpful throughout our revision process. The audio compact disc that comes with the text is the result of the tremendous effort of Dave Conrad, our engineer at Sound Concepts (Manhattan Beach, CA) and Chris Sterling, associate dean at George Washington University, who served as one of the voices on the CD. And once again, *Radio Production Worktext* is as good as it is because of the expert help and backing of the staff at Focal Press, especially Marie L. Lee, Senior Editor, and Tammy Harvey, Associate Acquisitions Editor.

CHAPTER ONE

The Production Studio

Information

1.1 INTRODUCTION

What may initially appear to be merely a roomful of electronic equipment will become a comfortable environment, once you've become familiar with the space and components that make up the production facility. The room that houses the equipment necessary for radio production work and in which a broadcaster's finished product is assembled is known as the **production studio**. If your facility has several studios, they may be labeled "Production 1" or "Prod. B" or maybe simply "PDX A." Most radio facilities have at least two studios; however, one is usually delegated as the **on-air studio** and is used for the live, day-to-day broadcasting. The other studio or studios are production studios, used for putting together programming material that is taped for playback at a later time. This includes such items as commercials, public-service announcements (PSAs), and station promotional announcements.

Some stations also have a studio that is considered a **performance studio**. It houses nothing more than microphones and table and chairs. The output is usually sent to a production studio to be taped, although sometimes it's sent directly to the on-air studio. This performance studio is used for taping interviews, for discussions involving several guests, or for putting a small musical group on the air.

Today most radio work is done **combo**; that is, the announcer is also the equipment operator. Because of this, the equipment and operator are in a single studio, be it a production room or an on-air room. In earlier radio days, the announcer was often located in a separate room or announce booth adjacent to the studio that housed the equipment. Visual contact and communication were maintained via a window between the two rooms. An engineer was required to actually manipulate the equipment, and all the announcer did was provide the voice. Many larger-market radio stations still use a similar announcer/engineer arrangement.

We should note here that anyone can be an announcer and work combo without a license. At one time, to be in charge of the station transmitter and make required station log entries, the operator needed a **Restricted Radiotelephone Operator Permit** issued by the Federal Communications Commission. FCC deregulation has eliminated many of the technical requirements of radio station operation and the need for this license. Another FCC license, the **General Radiotelephone Operator License**, is still available, but obtaining it requires a thorough knowledge of engineering practices and broadcast law, and the applicant must pass an FCC examination.

1.2 THE AUDIO CHAIN

Figure 1.1 shows a map of the typical radio production studio. Starting with various sound sources, such as an announcer's voice, a CD, or a tape, it shows the routes that sound takes to ultimately be broadcast or recorded. This is often called an **audio chain** because the various pieces of equipment are linked together. The trip can be complicated since the sound can go through several changes along the way. For example, it can be dubbed, or copied, from CD to cassette tape, or it can be **equalized**. The solid line shows sound being sent to the audio console, through **signal processing** equipment, and then to the transmitting system; this would be normal for an on-air studio. The broken line shows the sound being sent back to the various recorders after signal processing; this would be normal for a production studio. In both cases the sound can be heard in the studio through the monitor speaker. You'll learn more about all of this as you work your way through this text, but for now the diagram in Figure 1.1 provides a look at where you are headed.

The equipment shown is also representative of that found in the typical radio production studio. The **microphone** transforms the announcer's voice into an audio signal. It is not uncommon for a production facility to have one or more auxiliary mics for production work that requires two or more voices. Most production rooms also have two **CD players**, enabling different CDs to be played back-to-back or simultaneously. Generally these rooms still have two turntables so that records can be played in the same manner. Tape recorder/player sources include **reel-to-reel**, **cassette**, and **cartridge** machines. In modern studios, some of these machines will be **digital**, and the entire production process will be computer controlled. The num-

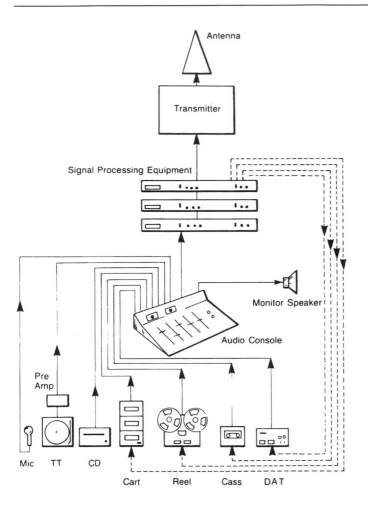

FIGURE 1.1   Production studio audio chain.

ber of recorders found in the production room depends on the complexity of the studio and the budget of the station.

All of this equipment feeds into the **audio console**, which allows the operator to manipulate the sound sources in various ways. Signal-processing equipment, such as an **equalizer**, **noise-reduction** system, or **reverb** unit, is usually put into the audio chain between the audio console and the transmitting equipment or recording equipment.

1.3 STUDIO LAYOUT

Almost all radio production studios use a U-shaped layout or some variation of it (see Figure 1.2). All the equipment needs to be within arm's reach of the operator, and the operator needs to be immediately in front of the audio console. With the use of remote start/stop switches for equipment that's out of convenient reach of the operator, all equipment manipulation occurs at the audio console once everything has been set up and cued. Compare Figures 1.1

and 1.2 to see how the audio chain translates into the actual production studio.

1.4 STUDIO CABINETS AND COUNTERS

Studio equipment is often installed on and in custom-built cabinets and counters. A less expensive but equally functional approach is to lay out the studio using modular stock components. Audio cabinets have been designed expressly for turntables, tape decks, audio consoles, and other pieces of studio equipment. For example, the studio configuration shown in Figure 1.2 is composed of modular units. Studios can be designed to fit the equipment a station has, and studios can even be expanded as new equipment is added. The current trend in studio furniture systems is to include space for computer monitors and other computer equipment that's being integrated with traditional broadcast equipment in the radio studio as shown in Figure 1.3. Both custom-built and modular cabinets and counters are also designed to provide easy access to the myriad cables neces-

FIGURE 1.2 U-shaped studio design. (Courtesy of the Express Group, San Diego, CA)

sary to wire all the studio equipment together yet maintain an attractive image for the look of the studio. Cabinets are also available for CDs, records, tapes, and other material that's kept in the production studio.

Does stylish furniture make a studio sound better? While that notion would be hard to quantify, a positive studio image does imply a commitment to high-quality production, and this often translates into more creativity, more productivity, and a better "sound" from that production studio.

1.5 SOUND SIGNAL VERSUS AUDIO SIGNAL

Much of the design of a production studio has to do with manipulating sound. It will be worthwhile to take a brief

FIGURE 1.3 Computer equipment integrated with production studio equipment. (Courtesy of Applied Construction Technology—KEZR-FM Production Studio)

look at sound since this will help you understand many aspects of the production process, whether it involves a sound signal or an audio signal. When sound is naturally produced (for example, an announcer speaking into a microphone), we think of that sound as a **sound signal**. In radio production, when that sound signal is manipulated electronically (such as recorded on audio tape), it's called an **audio signal**. Obviously most radio production must start at some point with a sound signal, but during the actual production process, we are often manipulating an audio signal. To further complicate things, these terms are often interchanged when people talk about various radio production processes.

1.6 SOUND DEFINED

When something vibrates, sound is generated. For example, plucking a single guitar string causes a mechanical vibration to occur, which we can easily see by looking at the string. Of course, we can also hear it. The vibrating string sets adjacent air molecules in motion, which in turn sets neighboring air molecules in motion and on and on. Sound develops waves (like a stone dropped into water), which vibrate up and down and set the air molecules in a push **(compression)** and pull **(rarefaction)** motion.

In addition to the vibration noted above, we need a medium for the sound to travel through. Of course, the medium we're usually concerned with is the atmosphere, or air. Sound can also travel through other materials, such as water or wood, but will often be distorted by the medium. Sound vibrations can't travel in a vacuum.

Finally, for sound to technically exist, we need a receiver. Someone (a person) or something (a microphone) must receive it and perceive it as sound.

Figure 1.4 shows a representation of sound being pro-duced. We can't actually see sound waves, but they act very much like water waves as we've noted. The sine wave (shown in Figure 1.5) is used to represent sound because it can readily show the wave compression (the portion of the wave above the center line) and the wave rarefaction (the portion of the wave below the center line).

1.7 CHARACTERISTICS OF SOUND WAVES

There are four key characteristics of sound that determine why one sound is different from another sound: amplitude, frequency, timbre, and the sound envelope. A sound wave's **amplitude** relates to its **volume**, or loudness. The loudness of a sound can be thought of as the height of the sound wave. The louder the sound, the higher the amplitude (see Figure 1.5A). As a sound gets louder, greater compression and rarefaction of air molecules takes place, and the crest of the wave will be higher while the trough of the wave will be deeper. A sound wave's amplitude is readily measured; however, loudness is a subjective concept. What is loud to one person isn't necessarily loud to another person.

Frequency relates to the **pitch** of a sound wave. The number of times per second that a sound wave vibrates (goes in an up and down cycle) determines its frequency, and how we hear these vibrations determines its pitch (see Figure 1.5B). The faster something vibrates, or the more cycles per second, the higher the pitch of the sound. Like amplitude, frequency can be objectively measured, but pitch is like volume in that it's subjective. The **wavelength** is the distance between two compressions (crests) or two rarefactions (troughs). The higher pitched sounds have the shorter wavelength.

In radio jargon, cycles per second are known as **hertz** (Hz). A sound wave that vibrates at two thousand cycles per second is said to have a frequency of 2000 hertz. When

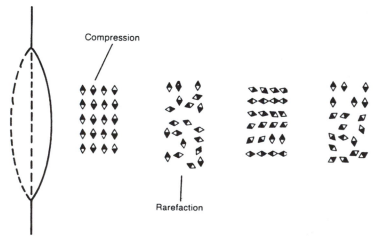

FIGURE 1.4 The production of sound.

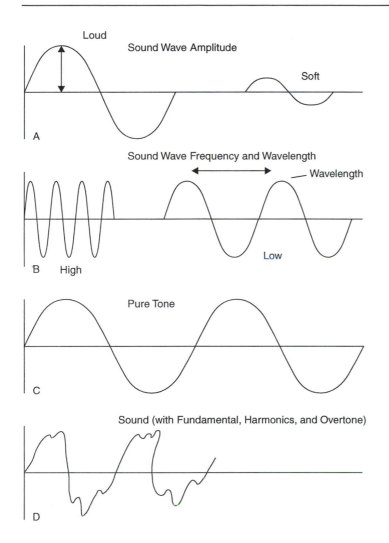

FIGURE 1.5 Characteristics of sound waves.

the cycles per second get higher, for example 20,000 hertz, the term **kilohertz** (kHz) is often used. It denotes 1000 cycles per second, so 20,000 hertz could also be called 20 kilohertz. A sound's **timbre**, or **tone**, relates to the **waveform** of the sound. It's the characteristic of sound that distinguishes one announcer's voice from another, even though both may be saying the same thing at the same volume and pitch. A graphic representation of a pure tone is shown as the shape of a sine wave, as in Figure 1.5C. Each sound has one basic tone that is its **fundamental**; most sound, however, is a combination of many tones with different strengths at different frequencies, so the waveform is much more complex, as shown in Figure 1.5D. These other pitches are either exact frequency multiples of the fundamental (known as **harmonics**) or pitches that are not exact multiples of the fundamental (known as **overtones**). For example, striking an A note (above middle C) on a piano would produce a fundamental tone of 440 Hz. In other words, the piano string is vibrating 440 times per second. The harmonics of this note will occur at exact (or whole number) multiples of the fundamental tone, such as 880 Hz (twice the fundamental) or 2200 Hz (five times the fundamental). The interaction of the fundamental, harmonics, and overtones creates the timbre of any particular sound. A sound's **wave envelope** relates to its **duration**, or the change in volume of a sound over a period of time. Normally a sound's wave envelope goes through four specific stages: **attack**, the time it takes an initial sound to build to maximum volume; **decay**, the time it takes the sound to go from peak volume to a sustained level; **sustain**, the time the sound holds its volume; and **release**, the time it takes a sound to die out from sustained volume to silence.

1.8 STUDIO SOUND CONSIDERATIONS

Several additional characteristics of sound need to be considered in designing the radio production studio. When sound strikes a surface (such as a studio wall), some of the sound is reflected back, some is absorbed within the material of the surface, and some is transmitted through the surface. Most of the sound that hits a hard, flat surface will be reflected back. However, if the surface is irregular, it will break up the sound wave and disperse the reflections—a phenomenon known as diffusion. Sound that's absorbed into the surface is dissipated within it. However, **penetration** occurs when sound goes through a surface and is transmitted into the space on the other side of the surface. Figure 1.6 illustrates that penetration, absorption, reflection, and diffusion are all characteristics that help determine the sound both produced and reproduced in the studio.

When a sound (such as an announcer's voice) is produced, the **direct sound** is the main sound that you hear. In a production situation, it is sound that goes from the announcer straight to the microphone. On the other hand, **reflected sound** reaches the microphone fractions of a second after the direct sound does because it has traveled an indirect route. Reflected sound consists of **echo** and **reverberation**. This indirect sound has bounced off or been reflected from one surface (echo) or two or more surfaces (reverb) before reaching the microphone (see Figure 1.7). Since it's an

early reflection, echo provides a distinct repetition of the sound, while reverb provides multiple echoes.

In designing the radio studio, the goal is to manipulate these sound considerations to create a proper sound environment for production work. When considering reflected sound, we think in terms of reverb ring and reverb route, with the same concepts being true for echo but to a lesser extent. **Reverb ring** or **reverb time** is the time that it takes for a sound to die out or go from full volume to silence—in other words, the decay—sustain-release portion of a wave envelope. **Reverb route** is the path that sound takes from its source to a reflective surface and back to the original source (or a microphone, if recording).

Excessive reflected sound tends to accent high and midrange frequencies, which produces a "harsh" sound; blur the stereo image, which produces a "muddy" sound; or cause standing waves (see section 1.10), which produces an "uneven" sound. Reflected sound can also be **reinforced** by causing objects or surfaces within the studio to vibrate at the same frequencies as the original sound in a sympathetic fashion.

Both **absorption** and diffusion are utilized to control reflected sound. Part of the reflected sound can be absorbed within the carpeting, curtains, and walls of the studio. Absorption soaks up sound and shortens reverb time to prevent excessive reflection. Excessive absorption provides a **dead studio**, which has a very short reverb ring and a long reverb route that produces a softer sound. In contrast, a **live studio** has a longer reverb ring and a short reverb route that produces a harder, or more brilliant, sound.

Diffusion uses irregular room surfaces to break up sound reflections. This decreases the intensity of the reflections, making them less noticeable, but doesn't deaden the sound as much because the sound reflects are redirected rather than soaked up. Most studio designs control reflections by a combination of absorption and diffusion techniques.

One common studio design is a **live end/dead end (LEDE)** approach. The front of the studio (where the announcer and equipment are located) is designed to absorb and diffuse sounds. This dead end quiets some of the equipment operation noise, picks up the direct sound of the announcer's voice, and absorbs the excess reflections that pass by the microphone from the live end. The live end, or back, of the studio adds a desirable sharpness to the sound by providing some reflected sound. This LEDE studio design was first devised by Synergetic Audio Concepts and has since been employed in many production studio situations.

1.9 STUDIO CONSTRUCTION MATERIALS

Part of these design considerations involve the actual construction materials used for the studio. Ideally, you want to keep penetration to a minimum by keeping outside (unwanted) sound from entering the studio and inside sound

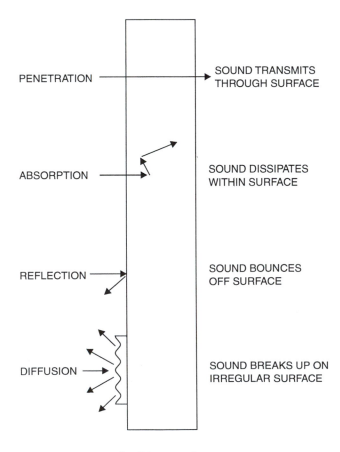

PENETRATION ——————————→ SOUND TRANSMITS THROUGH SURFACE

ABSORPTION ——————————→ SOUND DISSIPATES WITHIN SURFACE

REFLECTION ——————————→ SOUND BOUNCES OFF SURFACE

DIFFUSION ——————————→ SOUND BREAKS UP ON IRREGULAR SURFACE

FIGURE 1.6 Sound striking a surface.

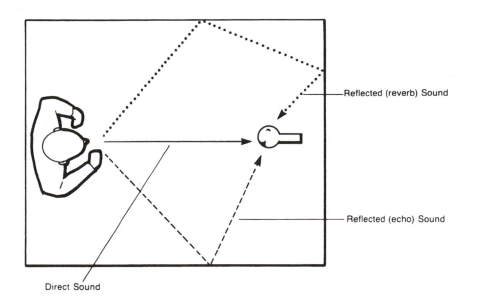

Reflected (reverb) Sound

Reflected (echo) Sound

Direct Sound

FIGURE 1.7 Direct versus reflected sound.

from escaping from the studio, except via the audio console. Radio studios utilize **soundproofing** to accomplish this. Doors are heavy-duty and tightly sealed, windows are usually double glass with the interior pane slanted downward to minimize reflected sounds, and walls, ceiling, and flooring use special sound-treatment materials. For example, studio walls may be covered with acoustically treated and designed panels that both absorb and trap reflected sounds (see Figure 1.8). Some stations use carpeting on the studio walls, and some production studios have actually used egg cartons on the walls as a sound treatment. (If you compare the design of an egg carton bottom with the design of the acoustic tile shown in Figure 1.8, you'll see

why some stations have gone the inexpensive egg carton route.)

All materials absorb sound to some degree, but each material will have a different **absorption coefficient**, which is the proportion of sound that it can absorb. A coefficient of 1.00 means all sound is absorbed in the material. On the other hand, a coefficient of 0.00 means no absorption, and all the sound is reflected back. Hard, smooth surfaces like plaster or panel walls and hardwood floors have low absorption coefficients. Heavy, plush carpets, drape-covered windows, and specially designed acoustic tiles will have higher coefficients. The purpose of soundproofing material is to give the studio a dead sound. Soundproofing absorbs and controls excess reverb and echo and produces a softer sound.

1.10 STUDIO SIZE AND SHAPE

The size and shape of a production studio can also determine how reflective the studio is. Any room that is highly reflective produces what is deemed a live sound. The radio production studio doesn't want to be overly reflective because the sound can be too bright and even harsh. Unfortunately, normal room construction often goes counter to good broadcast studio design. For example, studios with parallel walls (the normal box-shaped room) produce more reflected sound than irregularly shaped studios. Sound waves that are reflected back and forth within a limited area, such as between studio walls that are parallel, can produce standing waves. A **standing wave** is a combination of a sound wave going in one direction and an identical wave going in the opposite direction. This undesirable combined sound tends to be uneven, as previously mentioned. To

FIGURE 1.8 Sonex audio tiles. (Courtesy of Sonex Acoustical Division, Illbruck, Inc.)

help prevent standing waves, adjacent studio walls joined at irregular angles help break up reflected sound and control excessive reverb and echo.

The actual size of the production facility is partially determined by the equipment that must be housed in it. However, in constructing the radio production room, consideration should be given to the fact that when rooms are built with height, width, and length dimensions that are equal or exact multiples of each other, certain sound frequencies tend to be boosted, and certain sound frequencies tend to be canceled. Since this "peaks and valleys" sound is not desirable in the radio production room, cubic construction should be avoided when possible.

1.11 STUDIO AESTHETICS

There are some studio design considerations that can be categorized as the "aesthetics" of the production room. In general, the radio studio should be pleasant to work in; after all, the operator is confined to a rather small room for long stretches of time. For example, fluorescent lighting should be avoided. Not only does it tend to introduce hum into the audio chain, but it's harsher and more glaring than incandescent light. If possible, the studio lights should be on dimmers so that an appropriate level of light can be set for each individual operator.

Static electricity can be a problem in radio production studios because of the heavy use of carpeting. Most radio people don't enjoy getting shocked every time they touch the metal faceplate of a tape recorder. Also, some modern audio equipment has electronic circuits that can be disrupted by static discharges. If design factors can't keep the studio static free, commercial sprays can be put on the carpeting or spray fabric softener can be used to provide an antistatic treatment at a modest cost. A static touch pad could also be provided in the studio to keep static buildup at a minimum.

Stools or chairs used in the radio studio should be comfortable and functional. They must move easily because even though most of the equipment is situated close to the operator, he or she must move around to cue records, thread tapes, or select production music. The production stool must also be well constructed so that it doesn't constantly squeak if the operator moves slightly while the microphone is open. This may not be a factor for production studios designed for a stand-up operation, wherein there is no stool, and the counters are at a height appropriate for the operator to be standing while announcing. This allows the operator to be more animated in his or her vocal delivery and actually provides a better posture for speaking than a sitting position.

Many radio production rooms are decorated with music posters or radio station bumper stickers and paraphernalia. Not only does this keep the studio from being a cold, stark room, but it also gives the studio a radio atmosphere.

FIGURE 1.9 Studio on-air light. (Courtesy of Fidelipac Corporation)

1.12 ON-AIR/RECORDING LIGHTS

On-air lights (see Figure 1.9) are usually located outside the radio production room or studio. Normally they are wired so that whenever the microphone in that studio is turned on, the on-air light comes on. A light outside a production studio will often read "recording" instead of "on-air." In either case, a lit light indicates a live microphone. Good production practice dictates that when an on-air light is on, you *never* enter that studio, and if you're in the vicinity of the studio, you are quiet.

1.13 RADIO HAND SIGNALS

Radio **hand signals** don't play a major role in modern radio production; however, there are situations when vocal communication isn't possible, and hand signals are necessary. For example, if an announcer and engineer are working an on-air show from separate rooms, they must be able to communicate with each other. There are also times when two announcers must communicate, but an open or live microphone in the studio prevents them from doing so verbally. Because of situations like these, hand signals have evolved over the years to communicate some basic production information.

Often hand signals concern getting a program started or stopped. A **stand-by** signal is given just prior to going on-air by holding one hand above the head with the palm forward. The stand-by signal is immediately followed by the **cue talent** signal. Meaning, "You're on," this signal is given by pointing your index finger (using the same hand that gave the stand-by signal) at the person who's supposed to go on-air. The common hand signal for stopping a program is the **cut** signal, which is given by drawing the index finger across the throat in a "slitting" motion. This signal terminates whatever is happening at the moment and usually "kills" all live microphones.

Some hand signals are used to give directions to the announcer regarding the microphone. To get an announcer to

give mic level, for example, hold one hand in front of you with the palm down and thumb under the second and third fingers. Open and close the thumb and fingers in a "chattering" motion to indicate that the announcer should talk into the mic so that levels can be checked.

Other hand signals are often used during a production to let the talent know how things are going or to convey some necessary information. Timing cues are given with the fingers. To indicate that 2 minutes are left in the program, you would hold up the index and second finger of one hand in front of you. Using both index fingers to form a cross in front of you means there are 30 seconds left. Timing cues always tell the announcer how much time remains in the program. When everything is going fine, the radio hand signal is the traditional "thumbs up" given with clenched fist and extended thumb.

There is no universal set of hand signals, so you may find that your facility uses some that are different, uses some not presented here, or doesn't use any at all. In any case, an understanding of radio hand signals should prove helpful in some production situations. Figure 1.10 shows some of the basic radio hand signals.

FIGURE 1.10 Some basic audio hand signals.

1.14 NOISE AND DISTORTION

The next few chapters will deal with the equipment that's housed in the production studio. Inherent in any of this electronic equipment is noise. Any unwanted sound element introduced in the production process that was not present in the original sound signal is thought of as **noise**. For example, a microphone that employs an extremely long cable might add noise to the audio signal. Turntables and recorders can introduce noise from mechanical gears or just through the electronic circuits used in amplifying or recording the signal.

In broadcast production, the noise level should be kept as low as possible. Most radio production equipment is designed to produce a **signal-to-noise ratio (S/N)** of at least sixty-to-one. In other words, when 60 decibels of sound signal are reproduced by the equipment, only 1 **decibel (dB)** of noise is introduced. Obviously, the higher the S/N ratio of audio equipment the better.

Distortion is an unwanted change in the audio signal due to inaccurate reproduction of the sound. One type of distortion is loudness distortion, which can occur when a signal is recorded at a level too loud for the equipment to handle. An overdriven signal sounds "muddy," and the reproduced signal does not have the same clarity or sharpness that the original signal did.

You should be aware of noise and distortion when working with audio equipment, especially analog equipment. Digital equipment, in many instances, reduces the chance of introducing noise or distortion into your production work, but most of your production work will be accomplished with a combination of analog and digital equipment.

1.15 CONCLUSIONS

Unless you're building a radio facility from the ground up, it's probable that you will have little control over the construction of the studio; sound treatment is an important consideration, however, and some aspects of improving the sound environment can be put into practice in almost any situation. Completion of this chapter should have you in the radio production studio and ready to learn the procedures and techniques for operating all the equipment you see in front of you.

Self-Study

■ QUESTIONS

1. To work combo in radio means that the announcer _____.
   a) has an engineer to operate the studio equipment for him or her
   b) operates the studio equipment and also announces
   c) works at two different radio stations
   d) is announcing in both the on-air and the production studio

2. Which type of studio is least likely to contain an audio console?
   a) on-air studio
   b) production studio
   c) PDX studio
   d) performance studio

3. Sound produced in the radio studio that causes objects or surfaces within the studio to vibrate sympathetically is said to be _____.
   a) absorbed
   b) reflected
   c) reinforced
   d) diffused

4. In the radio production studio, sound that has bounced off one surface before reaching the microphone is _____.
   a) echo
   b) reverberation
   c) direct sound
   d) indirect sound

5. Reverb ring in the production studio refers to _____.
   a) the circular route reflected sound takes before it reaches the mic
   b) the time it takes reflected sound to go from full volume to silence
   c) just another common name for echo
   d) a sound that has bounced off two or more surfaces

6. The use of carpeting on the walls of some radio production facilities is an example of _____.
   a) an inexpensive way of decorating the studio
   b) producing reverb in the studio
   c) producing a live sound in the studio
   d) soundproofing the studio

7. Studios with parallel walls produce less reflected sound than irregularly shaped studios.
   a) true
   b) false

8. Most production studios use a U-shaped layout because _____.
   a) this places equipment within easy reach of the operator
   b) such a configuration must use incandescent lights rather than fluorescent lights
   c) this necessitates custom-built cabinets
   d) this uses the least amount of wire to connect the equipment

9. Static electricity is not a problem in the modern production studio because state-of-the-art audio equipment is impervious to static.
   a) true
   b) false

10. Which hand signal almost always follows immediately after the stand-by hand signal?
    a) 2 minutes to go
    b) thumbs up
    c) cue talent
    d) give mic level

11. If you hold up the index, second, and third fingers of one hand in front of you, the announcer knows _____.
    a) there are 3 minutes left in the program
    b) there are 30 seconds left in the program
    c) he or she should move three steps closer to the microphone
    d) three minutes have gone by since the beginning of the program

12. The linking of a CD player to an audio console, the console to an equalizer, and the equalizer to a reel-to-reel recorder would be called an _____.
    a) audio road map
    b) audio linking
    c) audio processor
    d) audio chain

13. When sound waves are reflected between parallel walls in such a manner that a wave reflected in one direction is combined with an identical wave going in the opposite direction, it produces an uneven sound known as a _____.
    a) diffused wave
    b) standing wave
    c) absorbed wave
    d) sympathetic wave

14. When a "recording" light is on outside a production studio, it means a microphone is "live" in that studio.
    a) true
    b) false

15. All studios utilize custom-built cabinets and counters to house the various pieces of equipment necessary for radio production.
    a) true
    b) false

16. When sound produced in the production studio strikes a hard, flat surface, which of the following does *not* happen?
    a) reflection
    b) absorption
    c) penetration
    d) diffusion

17. A production studio wall that has an absorption coefficient of .50 will absorb half the sound striking it and reflect back half the sound.
    a) true
    b) false

18. Posters and other radio station paraphernalia should not be put up in a production studio as it will distract the announcer from doing good production work.
    a) true
    b) false

19. Some of the acoustic tiles used in production studios use irregular surfaces to break up sound reflections. This is called _____.
    a) absorption
    b) reflection
    c) penetration
    d) diffusion

20. When sound is manipulated electronically (such as recorded onto audio tape), it is called a sound signal.
    a) true
    b) false

21. Which statement about "sound" is *not* true?
    a) sound is generated when something vibrates
    b) sound, to technically exist, must be heard
    c) sound vibrations develop waves by setting adjacent air molecules in motion
    d) sound vibrations travel faster in a vacuum than in air

22. An unwanted change in the audio signal due to inaccurate reproduction of the sound is _____.
    a) reverb
    b) noise
    c) distortion
    d) diffusion

23. Which of the following is *not* part of a sound wave's envelope?
    a) attack
    b) decay
    c) sustain
    d) rarefaction

24. The number of times a sound wave vibrates (goes in an up and down cycle) per second determines its _____.
    a) frequency
    b) amplitude
    c) wavelength
    d) wave envelope

25. As a final test of this chapter, match the items in the first list (1, 2, 3 . . .) with the choices in the second list (w, l, m . . .), and then select the correct set of answers from the sequences shown in a, b, c, or d below.

   1. _____ echo
   2. _____ direct sound
   3. _____ on-air studio
   4. _____ reverberation
   5. _____ absorbed sound
   6. _____ egg cartons
   7. _____ reinforced sound
   8. _____ cubic construction
   9. _____ live end/dead end
   10. _____ standing wave
   11. _____ frequency
   12. _____ wavelength
   13. _____ fundamental

   w. sound that goes into the walls of a studio
   l. used for live broadcasting
   m. sound that goes from the announcer to the mic
   o. sound bounced off one surface
   t. sound bounced off two or more surfaces
   v. sound that causes something in the studio to vibrate at its frequency
   s. used for soundproofing
   h. rooms with height, width, and length dimensions that are the same
   d. a combination of identical sound waves going in opposite directions
   a. a studio design in which one end of the studio tends to absorb sounds and the other end reflects sounds
   u. the basic tone that each sound has
   n. the number of times a sound wave goes up and down per second
   b. the distance between two crests or troughs of a sound wave

   a) 1.t 2.w 3.l 4.o 5.v 6.s 7.m 8.h 9.a 10.d 11.u 12.n 13.b
   b) 1.o 2.m 3.l 4.t 5.w 6.s 7.v 8.h 9.a 10.d 11.n 12.b 13.u
   c) 1.t 2.h 3.a 4.o 5.w 6.l 7.v 8.m 9.d 10.s 11.b 12.u 13.n
   d) 1.t 2.m 3.l 4.o 5.w 6.s 7.v 8.h 9.a 10.d 11.n 12.b 13.u

■ ANSWERS

If You Answered A:

1a. No. This is not working combo. (Reread 1.1.)
2a. No. An on-air studio needs one for sending the mixed signal out. (Reread 1.1 and 1.2.)
3a. No. If anything, absorbed sound would be diminished. (Reread 1.8.)
4a. Yes. Echo is sound that has reflected off a single surface.
5a. No. This is close to (but not exactly) describing reverb route. (Reread 1.8.)
6a. No. Painted walls would be less expensive, so this can't be correct. (Reread 1.9 and 1.11.)
7a. No. Just the opposite is true. (Reread 1.10.)
8a. Right. The operator can reach around the "horseshoe."
9a. No. Just the opposite is true. (Reread 1.11.)
10a. Wrong. The two-minute signal would not come until the end of a program. It won't be right after a stand-by signal. (Reread 1.13.)
11a. Yes. This is the correct hand signal.
12a. No. While you could map out this audio signal route, this is not the best answer. (Reread 1.2.)
13a. No. Diffused waves would be sound reflections that have been broken up. (Reread 1.8 and 1.10.)
14a. Right. On-air and recording lights usually come on automatically when a microphone is on in that studio.
15a. No. While some do, many stations use modular, stock components, or homemade cabinets and counters. (Reread 1.4.)

16a. No. Most of the sound that strikes a hard, flat surface will be reflected. (Reread 1.8.)

17a. Yes. This would be a true statement.

18a. Wrong. Posters and other studio decorations aren't likely to cause poor production work, and they do add "atmosphere" to the studio. (Reread 1.11.)

19a. No. This would be a "soaking up" of sound reflections. (Reread 1.8 and 1.9.)

20a. Wrong. A sound signal is what the original sound is. (Reread 1.5.)

21a. No. This is a true statement. (Reread 1.6.)

22a. No. Reverb is a form of reflected sound. (Reread 1.8 and 1.14.)

23a. Wrong. Attack is the time it takes an initial sound to build up to full volume. (Reread 1.7.)

24a. This is the correct answer.

25a. No. You are confused about different kinds of sound. (Reread 1.8.)

If You Answered B:

1b. Correct. The announcer is also the equipment operator when working combo.

2b. No. A production studio needs one for mixing. (Reread 1.1 and 1.2.)

3b. Wrong. Reflected sound is sound that has bounced off a surface. (Reread 1.8.)

4b. No. Reverb is sound that has reflected off two or more surfaces. (Reread 1.8.)

5b. Yes. This is what we call reverb ring.

6b. Wrong. Carpeting absorbs sound and reduces reverb. (Reread 1.8 and 1.9.)

7b. Yes. This is the correct response.

8b. No. Lights have no relevance, so this can't be correct. (Reread 1.3 and 1.11.)

9b. Yes. Modern electronics, especially logic circuits, can be disrupted by static discharges.

10b. Wrong. You wouldn't know the production was going well if it hadn't started yet. (Reread 1.13.)

11b. No. Crossed index fingers indicate 30 seconds. (Reread 1.13.)

12b. No. This is not correct. (Reread 1.2.)

13b. Correct. This is the right answer.

14b. Wrong. This is exactly what it means. (Reread 1.12.)

15b. This is the correct response.

16b. This isn't a bad choice, but some sound will be absorbed and dissipated even with hard surfaces. (Reread 1.8.)

17b. No. A coefficient of 1.00 would mean total absorption and a coefficient of 0.00 would mean no absorption. (Reread 1.9.)

18b. This is the correct answer.

19b. No. This would be sound that has bounced off a surface. (Reread 1.8.)

20b. Yes. This would be an audio signal, so this statement is false.

21b. No. This is a true statement. (Reread 1.6.)

22b. No. You're close because noise is an unwanted element introduced into the audio signal that was not present in the original sound, but there's a better response. (Reread 1.14.)

23b. Wrong. Decay is the time it takes sound to go from peak volume to a sustain level. (Reread 1.7.)

24b. Wrong. Amplitude relates to volume and the height of a sound wave. (Reread 1.7.)

25b. Right. You have now finished this chapter.

If You Answered C:

1c. No. This is not working combo. (Reread 1.1.)

2c. No. This is just another term for production studio. (Reread 1.1 and 1.2.)

3c. Right. This is the correct response.

4c. No. Direct sound doesn't bounce off any surface before reaching the mic. (Reread 1.8.)

5c. No. Echo and reverb are both reflected sound but distinctly different. (Reread 1.8.)

6c. No. Just the opposite would happen. Soundproofing with carpeting would help produce a dead sound in the studio. (Reread 1.8 and 1.9.)

8c. No. In a cost-minded facility, this could be a negative. (Reread 1.3 and 1.4.)

10c. Correct. Stand-by and cue talent hand signals are always given one after the other.

11c. No. There is another hand signal to move the announcer closer to the mic, and exact steps are never indicated. (Reread 1.13.)

12c. No. You are way off base with this answer. (Reread 1.2.)

13c. No. Absorbed waves would be sound reflections that have been soaked up. (Reread 1.8–1.10.)

16c. Wrong. Some sound will penetrate a hard surface and be transmitted to the adjoining space. (Reread 1.8.)

19c. No. This would be sound that has been transmitted through a surface. (Reread 1.8.)

21c. No. This is a true statement. (Reread 1.6.)
22c. Yes. You're correct.
23c. Wrong. Sustain is the time a sound holds its volume. (Reread 1.7.)
24c. Wrong. Wavelength refers to the distance between two wave compressions or rarefactions. (Reread 1.7.)
25c. No. You're confused about a great number of things. (Reread the entire chapter.)

If You Answered D:

1d. No. This seems improbable and is not working combo. (Reread 1.1.)
2d. Correct. It usually only has mics that are fed to an audio console in either a production studio or an on-air studio.
3d. Wrong. (Reread 1.8.)
4d. You're partly right, but echo and reverb are both indirect sound, and one is a better response to this question. (Reread 1.8.)
5d. No. While this describes reverb, there is a better response. (Reread 1.8.)
6d. Correct. Carpeting walls helps to soundproof, as would use of acoustic tiles designed for the production studio.
8d. While this may be true, it really is not the best reason. (Reread 1.3.)
10d. No. The give-mic-level signal, if used, would have been given before a stand-by signal. (Reread 1.13.)
11d. Wrong. Time signals are usually given only to show how much time remains in a program. (Reread 1.13.)
12d. Correct. The term "audio chain" describes how broadcast equipment is connected together.
13d. No. Sympathetic waves would be sound reflections that have been reinforced. (Reread 1.8 and 1.10.)
16d. Right. Sound is diffused when it strikes an irregular surface.
19d. Correct. This is diffusion.
21d. Yes. Sound vibrations can't travel in a vacuum.
22d. No. Diffusion is sound that has been "broken up" by an irregular surface. (Reread 1.8 and 1.14.)
23d. Correct. Rarefaction is not part of a sound's wave envelope.
24d. Wrong. Wave envelope refers to a sound's duration. (Reread 1.7.)
25d. You're almost right, but not quite. (Reread 1.8.)

Projects

■ PROJECT 1

Tour a radio station and write a report describing its production facilities.

Purpose

To enable you to see a commercial radio production facility firsthand.

Advice, Cautions, and Background

1. Don't push a station that seems reluctant to have you come. Some stations (especially smaller ones) are happy to have you. Others are pestered to death with would-be visitors or aren't equipped to handle them.
2. Make sure that before you go, you have some ideas about what you want to find out so that you can make the most of your tour.
3. Keep your appointment. Once you make it, don't change it, for this will breed ill will for you and your school.

How to Do the Project

1. Select a station that you would like to tour. (If the instructor has arranged a station tour for the whole class, skip to step 4.)
2. Call the station, tell them you would like to see that station so you can write a report for a radio production class, and ask if you may come.
3. If they're agreeable, set a date; if they're not, call a different station.

4. Think of some things you want to find out for your report. For example:
   a. How many production studios do they have?
   b. What types of equipment (CD player, reel-to-reel recorder, etc.) do they have?
   c. What manufacturers (brand names) have they bought equipment from?
   d. How is the production studio soundproofed?
   e. Is the on-air studio different from the production studio(s)?
   f. Do the announcers ever use hand signals during a production?
   g. Are their studios designed for stand-up operation?
   h. What is the physical layout of the studios and the station?
5. Go to the station. Tour to the extent that they'll let you, and ask as many questions as you can.
6. Immediately after leaving the station, jot down notes so you'll remember main points.
7. Write your report in an organized fashion, including a complete description of the production studio and the other points you consider most pertinent. It should be two or three typed pages. Write your name and RADIO FACILITY TOUR on a title page.
8. Give the report to your instructor for credit for this project.

◼ PROJECT 2

Redesign your production studio.

Purpose

To suggest improvements to your production facility, utilizing some of the techniques mentioned in this chapter.

Advice, Cautions, and Background

1. While you may initially feel your production studio is "perfect" just the way it is, almost every studio can be configured better.
2. You won't be judged on artistic ability, but make your drawings as clear as possible.
3. You may find it useful to complete Project 1 before attempting this project.

How to Do the Project

1. Draw a rough sketch of your production studio, showing approximate dimensions, door and window locations, equipment placement, etc.
2. Draw another sketch of the studio, suggesting changes or improvements to it. For example, if there currently is a CD player on the left side of the audio console and another on the right side, you may suggest moving them both to one side. If you notice a paneled or painted sheet rock wall in the studio, you may suggest putting acoustic tile on that area. You may want to employ an idea for your studio that you noticed when you did Project 1, but be creative, and try to design the best possible production studio.
3. On a separate sheet of paper, provide a reason for each change you suggest.
4. Write your name and STUDIO DESIGN on a title page, and put your two sketches and reasons together.
5. Turn in this packet to the instructor to receive credit for this project.

◼ PROJECT 3

Draw an audio chain flowchart for your production studio.

Purpose

To help you understand that the audio chain maps the route an electronic audio signal takes as it goes from one place to another in the production studio.

Advice, Cautions, and Background

1. It may be helpful to review Figure 1.1 before beginning this project.
2. Use simple shapes to represent equipment and arrowed lines to represent the sound signal.
3. You won't be judged on artistic ability, but make your drawings as clear as possible.

How to Do the Project

1. Pick a single sound source in your production studio, such as a CD player.
2. Draw a figure to represent the CD player toward the left side of a sheet of paper, and label it appropriately.
3. Determine where the sound goes as it leaves the output of the CD player. (Most likely, to the audio console.)
4. Draw a figure to the right of the CD player to represent the audio console, and label it.
5. Draw an arrowed line going from the CD player to the audio console to represent the signal flow.
6. Now determine where the sound goes next. (It could go to a signal processor or maybe directly to an audio tape recorder.)
7. Continue in this manner until you've drawn all the possible signal paths that the CD player sound could take. (Don't forget to include the signal to the studio monitors!)
8. Pick another sound source, such as the studio microphone, and repeat the above steps. Do the same for all the other sound sources in your studio—audio tape recorders, turntables, etc.
9. Write your name and AUDIO CHAIN on your sketch, and turn it in to the instructor to receive credit for this project.

CHAPTER TWO

The Audio Console

Information

2.1 INTRODUCTION

The audio console, or **control board**, is the primary piece of equipment in any production facility. It can be more difficult to understand than other pieces of equipment in the radio production studio, but most other pieces of equipment operate through the audio console. Therefore, unless you can operate the audio console, you can't really utilize other studio equipment, such as a compact disc player or audio tape recorder. An audio console is somewhat like the receiver or amplifier that your home stereo system employs. To use your record player or CD, you must have it plugged into that receiver.

Broadcast, or on-air, consoles used in most radio stations are fairly straightforward in their construction and operation; however, some control boards used in audio production and music recording become much more complex with additional controls and features. Regardless, all control boards have basic similarities. Even though you'll run across many different brands of audio consoles in your radio production work, a thorough knowledge of any one control board will enable you to use any control board after a brief orientation.

Like a lot of other audio production equipment, the audio console is facing transformation from analog to digital. In the future, rather than having a physical audio board containing individual input and output modules with multiple controls, the operator may simply manipulate a "virtual console," or representation of an audio console on a computer screen. This is already common practice on some digital audio workstations that include mixing functions (see Figure 2.1). A more detailed look at digital equipment will take place in Chapter Ten, but for now let's look at the traditional audio console that is still a mainstay of most production facilities.

Look at the audio console shown in Figure 2.2. Most boards, from the simplest to the most complex, include some method for input selection (mic/line switch, A/B/C source selectors) and input volume control (faders and pots) as well as some method for output selection (program/audition/auxiliary switch, output selectors) and volume control (master gain control). They should also have some method for indicating to the operator the strength of the signal (VU meter) and a way of allowing the operator to hear the mix of sources (monitor speakers, cue speakers, headphones). Boards also have amplifiers at various stages so that the signal is loud enough when it eventually goes to the transmitter or an audio tape recorder. These amplifiers are buried inside the board and are not something the board operator can control. In addition, audio consoles can have many other special features to help the board operator work more efficiently and creatively. The board may look intimidating because of all the buttons, knobs, and levers, but most of them are repeats of each other since the board has many different inputs and outputs. These will be explained in detail as we begin to explore the operation of the audio console.

2.2 AUDIO CONSOLE FUNCTIONS

The control board has three primary functions: to mix, to amplify, and to route. First, the audio console enables the operator to select any one or a combination of various inputs. In other words, it must first be determined where the signal is coming from: microphones, CDs, turntables, or tape cartridge players, for example. Audio consoles are sometimes referred to as mixing boards because of their ability to select and have several inputs operational at the same time. Much production work will be a mix of voice, music, and sound effects.

The second function of the control board is to **amplify** the incoming audio signal to an appropriate level. Most sound sources (such as a microphone or turntable) produce a small electrical current that must be amplified to be used. The third function of the audio console is to enable the operator to route these inputs to a number of outputs, such as monitor speakers, the transmitter, or an audio tape recorder. This allows us to determine where the signal is going and to provide a means for listening to the signal.

2.3 BASIC AUDIO CONSOLE PARTS

All control boards operate in basically the same way. For purposes of simplicity, let's assume the board is a very

FIGURE 2.1 "Virtual" audio console controls. (Courtesy of Innovative Quality Software)

small monaural audio console with two inputs (one for a microphone and one for a CD player) and one output, which goes to an audio tape recorder. Look at Figure 2.3A. This is considered a two-channel board with the microphone assigned to channel 1 and a CD player assigned to channel 2. In general terms, a channel refers to the path an audio signal follows. On an audio console, a **channel** refers to a group of switches, faders, and knobs that are usually associated with one sound source. On this board, note the individual input selectors, output selectors, and input volume controls associated with channel 1 and channel 2. The output gain control, VU meter, and monitor gain control are associated with both channels.

2.4 INPUT SELECTORS

The **input selectors** on this particular model of audio console are push buttons that can be put in either the mic or the line position. The reason for the two positions is that different pieces of equipment are amplified differing amounts. Microphones generally do not have amplifiers built into them, whereas CD players and audio tape recorders have already put their signals through a small amount of ampli-

fication. When the input switch is in the mic position, it sends the signal coming into it through a stage of preamplification that is not present for signals coming into the line position. In other words, the mic position allows a signal to catch up to a signal coming into the line position in terms of amplification. Then they both often go through additional preamplification.

The way the input selector switches are arranged in Figure 2.3A, the microphone comes into the first (left) input, and the CD player comes into the second input. This means the mic has to be patched into the first input and the CD player into the second input. This patching involves running a cable from the mic and CD player to the back (or bottom) of the audio console. Such wiring is usually done in a semipermanent way by the engineer (see Figure 2.3B).

The way this console is designed, the microphone could be coming into the second input and the CD player into the first. The patching would need to be changed (as shown by the dotted lines in Figure 2.3B). The input selector switches on the front of the audio console would also need to be reversed.

The mic and the CD player could also both be patched into the first input. Then when the input selector switch is pushed in the mic position, the mic would be activated, and when it's in the line position, the CD player would be activated. Under this configuration, however, the mic and CD player could not be used at the same time. Not all audio boards have input selector switches. Some radio production boards have certain inputs that must be at the mic level and other inputs that can only accommodate equipment that has been preamplified and is ready for a line level. On boards of this type, usually only microphones can be patched into the first two inputs, and only CD players, tape recorders, and other line-level equipment can be patched into the remaining inputs.

On the other hand, some boards have input selector

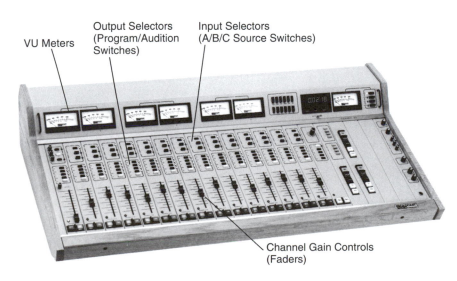

FIGURE 2.2 Audio console. (Courtesy of Fidelipac Corporation)

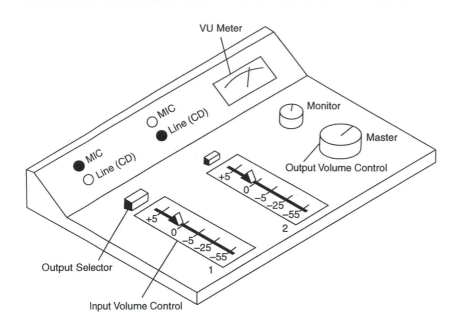

FIGURE 2.3A Simplified audio console.

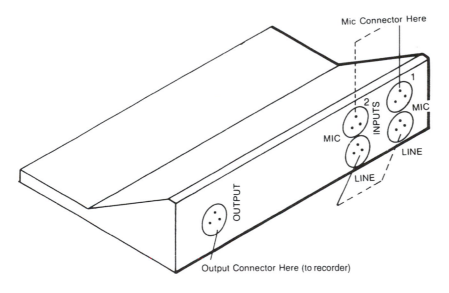

FIGURE 2.3B Rear of simplified audio console.

switches that have three or more positions for one input. For these boards, it's possible to patch a CD player at position A, a turntable at position B, and a cartridge tape player at position C all into the same input. The use that the facility was going to make of the various equipment would have to be carefully studied because, of course, no two pieces of equipment could be used at the same time on a single channel.

Regardless of the configuration of an audio board, the first two channels (from the left) are almost always utilized as mic level channels. Channel 1 is normally the main studio mic, and channel 2 is often an auxiliary microphone.

2.5 INPUT GAIN CONTROL

The input **gain controls** shown in Figures 2.2 and 2.3A are called **sliders**, or **faders**. They are merely **variable resistors**. Although they are called **volume controls**, or **gain controls**, they don't really vary the amount of amplification of the signal. The amplifier is always on at a constant volume. Raising the fader (moving it from a "south" to a "north" position) decreases the amount of resistance to this signal. When the fader is raised and the resistance is low, a great deal of the signal gets through. The dynamic is like that of a water faucet. The water volume reaching the faucet is always the same, even when the faucet is closed. When you open the faucet (decrease the resistance), you allow the water to flow, and you can vary that flow from a trickle to a steady flow.

Some boards have rotary knobs called **potentiometers**, or **pots**, instead of faders. These provide the same function. As the knob is turned to the right (clockwise), the resistance is decreased and the volume is increased. Some production people feel the fader is easier to work with. For one thing, the fader gives a quick visual indication of which channels are on and at what level. This is harder to see with a rotary knob.

The numbers on both rotary knobs and faders may be in reverse order on some audio consoles to show their relationship to resistance. For example, if a knob is completely counterclockwise, or off, it may read 40; at a twelve o'clock position it may read 25; and completely clockwise, it may read 0. These figures represent decreasing amounts of resistance and thus higher volume as the knob is turned clockwise. Modern boards with fader volume controls often use a range of numbers from –55 to 0 to +10 or +15. While the same relationship to resistance is true (the more the fader is raised, the less resistance), these numbers actually relate to decibels and the VU meter. If the board has been set up properly, a 0 setting of the fader will produce a 0 reading on the VU meter. Some boards avoid using any numbers at all and merely use equally spaced indicator lines to provide some kind of reference for various knob and fader settings.

Of course, most boards have more than the two inputs of our example board. In most radio production studios, boards have 10 to 12 channels. In professional audio production facilities, 16 or 20 inputs is not uncommon. Each input has its own gain control.

In addition to the gain controls just mentioned, some boards have a **gain trim**, or **trim control**, that fine-tunes the volume of each input. For example, if the sound signal coming from a CD player has the left channel louder (stronger) than the right channel, the trim control can decrease the left level or increase the right level until the sound signal is equal for both channels. Each input channel on an audio console usually has a gain trim feature, and often there is a similar trim adjustment for the program and audition output of the audio board. While these trim controls may be on the face of the audio board, they're often also an internal adjustment that is taken care of by the engineer when the control board is initally set up.

2.6 MONITORING

Once the signal is through the input gain controls, it's amplified in a program amplifier and then sent several places (see Figure 2.4). One of these is a **monitor amplifier**. This amplifies the signal so that it can be sent into a **monitor speaker** to enable the operator to hear the signal that is going out. Boards usually contain a simple potentiometer to control the gain of the monitor speaker. This control in no way affects the volume of the sound being sent out to the

audio tape recorder (or transmitter, etc.). It only controls the volume for the person listening to it in the control room. A common mistake of beginning broadcasters is to run the studio monitors quite loud and think all is well, while in reality they have the signal going through the audio board (and therefore to a recorder or on-the-air) at a very low level. It's important for the operator to be aware of the level of sound going out the line.

2.7 CUE

Another function found on most boards is called **cue**, which allows you to preview an input. Both rotary pots and fader controls go into a cue position, which is below the off position for that control. If you turn the rotary pot all the way counterclockwise to off, it will reach a detent, or stop. Keep turning the knob (with a little extra pressure) until it clicks into the cue position. Faders are brought down, or "south," until they click into cue (see Figure 2.5). Some faders can be put into cue with a separate pushbutton (see Figure 2.8) that, when depressed, puts that channel into cue regardless of where the fader control is set. Cue position is usually marked on the face of the audio console.

In the cue position, the audio signal is routed to a cue amplifier and then to a small speaker built into the control board. Since the quality of this small internal speaker is usually marginal at best, the cue signal is often sent to a small, but better quality, external speaker located near the audio console. Some audio consoles send the cue signal to the main

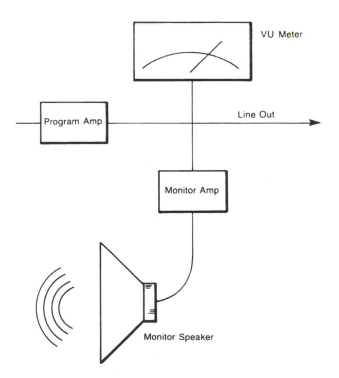

FIGURE 2.4　Monitor amp section of audio console.

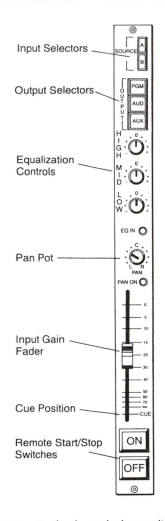

Input Selectors

Output Selectors

Equalization
Controls

Pan Pot

Input Gain
Fader

Cue Position

Remote Start/Stop
Switches

FIGURE 2.5 Single channel of an audio console.

studio monitor speakers. The program signal is automatically turned down, or "dimmed," when a channel is put into cue and the cue signal is heard "on top of the program signal."

As the name implies, this position is designed to allow the operator to cue up the sound source. For example, a record can be cued to the exact beginning so that the sound will start immediately when the turntable is turned on. If an input is in cue, the signal doesn't go to the tape recorder or to any other output source such as the transmitter. Its only purpose is to allow off-air cueing.

Many beginning announcers and production people forget to move the volume control out of the cue position after cueing up the sound source. If something is left in cue, it won't go out on the air or be routed to an audio tape recorder. It will only play through the cue speaker.

2.8 HEADPHONE USE

Most audio consoles also have provision for listening to the output of the board through **headphones**. Since live mi-

crophones are often used in production work, the monitor speakers are muted when the mic is on so that **feedback** doesn't occur. To be able to hear an additional sound source, such as a CD, headphones are necessary. Audio consoles allow you to monitor any of the outputs with headphones by selecting an appropriate switch. There is usually a volume control to adjust the signal level going to the headphones.

2.9 VU METERS

Another place the signal is sent after program amplification is the volume unit indicator, or **VU meter** (see Figure 2.6). This is a metering device that enables the operator to determine what level of sound is going out the line.

One common type of VU meter has a moving needle on a graduated scale. Usually the top position of the scale is calibrated in **decibels** (**dB**), and the lower portion of the scale is calibrated in percentage. In audio engineering, a reading of 0 dB is 100 percent volume, or the loudest you want the signal to go. The VU meter is important for consistent audio production work. As noted in Chapter One, how loud something sounds is very subjective. What is loud to one announcer may not be deemed loud by another, especially if they set the monitor speaker volume differently. The meter gives an electronic reading of volume that is not subjective.

The accuracy of VU meters is sometimes questioned in two areas. First, VU meters have trouble indicating transients, or sudden sharp, short increases in volume of the sound signal. Most VU meters are designed to indicate an average volume level and ignore these occasional sound bursts. Secondly, VU meters tend to overreact to the bass portion of the sound. In other words, if a sound signal is heavy in the bass frequencies, it will probably indicate a higher VU reading than the total sound signal is actually providing. In spite of these concerns, the VU meter remains the best indicator of volume levels in broadcast production.

Generally, an operator should control the signal so that it stays approximately between 80 percent and 100 percent. When the needle swings above 100 percent, we say the signal is **peaking in the red** because that portion of the VU

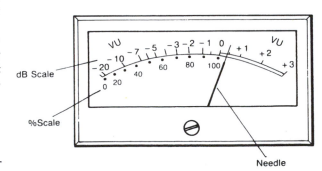

FIGURE 2.6 Standard VU meter.

meter usually has a red line. This is an indication to the operator to lower the gain (increase the resistance) of the fader or pot. Occasional dips into the red portion of the scale are likely, but having the needle consistently above 0 dB should be avoided.

Users of digital production equipment will quickly find that, unlike analog equipment, digital equipment is very unforgiving in regards to recording "in the red." Most digital equipment will not tolerate recording at any level above 100 percent and will distort or add "pops" to any recorded signal that exceeds it. Good production practice says to record everything at −10db when recording on digital equipment.

Often at the end of the needle's movement is a metal peg to prevent the needle from going off the dial. Allowing the gain to become so high that the needle reaches the upper peg is called **pegging the meter**, or **pinning the needle**, and should be avoided to prevent damage to the meter as well as distortion of the signal.

When the signal falls below 20 percent consistently, we say the signal is **riding in the mud**, and the operator should increase the volume. If it's necessary to adjust the level during the program, we say that the operator is **riding the gain**. Gain is an audio engineer's term for loudness or volume. This is why a radio operator is often called a disc jockey. He plays record *discs* and *rides* the gain.

Our simple audio board has only one VU meter that indicates the volume of the sound going out (review Figure 2.3A). Many boards have multiple VU meters. For example, they might have separate meters for each input. On our simple board, if the meter is peaking in the red, we would not know just by looking at the VU meter if the mic or the CD player is the culprit. But having separate meters for the mic and CD player would show which had the high volume. Boards also have multiple meters if they have multiple outputs. For example, a board that is stereo will have one meter for the right channel and one for the left.

On some boards the VU meter isn't an electromagnetic meter as has been described so far, but a succession of digital lights (LEDs, or light-emitting diodes) that indicate how high the volume is. Other electronic meters replace the LEDs with liquid crystal or fluorescent displays, and the advantage of these meters over mechanical meters is that they can indicate volume changes more quickly and accurately (see Figure 2.7).

2.10 OUTPUT SELECTORS

Another place that sound could go after program amplification is to an audio tape recorder. On the board in Figure 2.3A, the output selectors are just on/off buttons because the only place that the signal is intended to go is the recorder. However, the configuration of **output selector** buttons varies from board to board. If there are a large number of outputs—six, for example—then there may be six output buttons for each input. The input signal will be sent to whichever buttons are down, or selected. For instance, the microphone could be sent to the transmitter; the turntable, to a reel-to-reel recorder; and the CD, to a cassette recorder. Sometimes there are no output selectors; every input either goes out or it doesn't, depending on the master volume control. Other boards have buttons labeled "send" that determine where the signal goes.

The most common arrangement for the output selectors on a radio production board is a bank of three buttons for **program**, **audition**, and **auxiliary** outputs. When no button is pushed in, the output is stopped at this point. When the program button is pushed, the signal goes to the transmitter or to a tape recorder. The program position would be the normal operating position when using an audio console.

If the switch is in the audition position, the signal is sent to an audition amplifier and then out to monitor speakers, tape recorder, and so on. The signal will *not* be sent to the transmitter. The purpose of the audition switch is to allow off-air recording and previewing of the sound quality of a particular signal. For example, you could be playing a CD on one CD player through channel 3 of the audio console (in the program position) and at the same time be previewing another CD on the other CD player through channel 4 in the audition position. Each channel of the audio console can be used either in the program or audition position.

The auxiliary, or **aux**, position is just like either the program or audition position and is another output for the audio board. For instance, some studios are set up so that the aux position of one control board feeds another studio and becomes an input on the audio console in the other studio. Unlike most input selectors, which allow only one input to be selected at a time, output selectors usually allow more than one output to be active at a time. If all output buttons were depressed, you could send the same audio signal to three different locations at the same time.

Some boards just use a single toggle switch in place of push buttons. The functions are exactly the same, depending on which way the toggle is switched; if the toggle is in the middle, or neutral, position, the signal stops at that point (just like having no button pushed in a three-button bank).

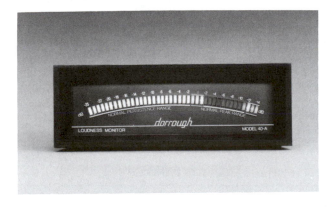

FIGURE 2.7 VU meter (fluorescent). (Courtesy of Dorrough Electronics)

2.11 OUTPUT GAIN CONTROL

The only output gain control on our simple board in Figure 2.3A is called the **master fader** because the signal is intended to go only one place. If both output select buttons are in the down (on) position and the master fader is up, the signal will go to the recorder. It is, of course, possible to send the microphone but not the CD player signal or vice versa. Also, if both buttons are on but the master fader is all the way down, the signal won't be recorded.

Many boards have more than one output gain control. Again, a stereo board requires two masters, one for the right channel and one for the left. Board operators often want to record what they're sending to the transmitter, so they need additional outputs to go to an audio tape recorder. If there are a large number of output volume controls, then there's usually a master gain control that overrides all the other output volume controls. In other words, if the master is down, the signal won't go anywhere, even though one or more of the output volume controls are up.

Boards are often referred to by their number of inputs and outputs. A six-in/four-out board has six inputs and four outputs.

2.12 REMOTE STARTS, CLOCKS, AND TIMERS

Other "bells and whistles" that frequently appear on audio boards include **remote start switches** (see Figure 2.5). These are usually located below each individual channel gain control, and if the equipment patched into that channel (such as a CD player or audio tape recorder) has the right interface, it can be turned on, or started, by depressing the remote start. This makes it easier for the announcer to start a record while talking into the microphone without having to reach off to the side and possibly be pulled off mic. For most consoles, these switches also turn that channel on and off. In other words, even if you have the fader for that channel turned up, no sound signal will go through the channel until it has been turned on. Some consoles are set up so that they will automatically turn the channel on when the fader is moved upward.

Many control boards now include built-in clocks and **timers**. Digital clocks conveniently show the announcer the current time (hours, minutes, and seconds), and timers can be reset at the start of a CD or tape to count up the elapsed time or count down the remaining time. Many timers will begin automatically when an "on" or remote start button is pressed.

2.13 FREQUENCY RESPONSE

In radio production, we often mention the frequency response of equipment or, for that matter, the frequency response range of human hearing. In very general terms, we can think of the human ear as able to hear frequencies within the range of 20 to 20,000 cycles per second. For most of us, it's not quite so low nor quite so high. In any case, radio production equipment, such as an audio console, should be able to reproduce an audio signal in that range, and most modern equipment is measured by how well it does so. For example, a monitor speaker may have a frequency response of 40 hertz to 18 kilohertz, meaning that that speaker can accurately reproduce all frequencies within that range. An inexpensive broadcast microphone may have a frequency response of only eighty hertz to thirteen kilohertz. It would not be able to pick up any of the higher frequencies, above 13,000 hertz. This would not be a problem if the mic were used primarily to record speech because the human voice usually falls in a frequency range of 200 to 3000 hertz. Obviously, if you wanted to record a musical group (which often has sounds in the full range of frequencies), you would want to use a microphone with a wider frequency response. Frequency response is often shown with a frequency response curve because some equipment may not pick up or reproduce some frequencies as well as others. Since most broadcast equipment is designed to pick up all frequencies equally well, its response is considered to be **flat**, although few components have a truly flat frequency response curve.

While there are no standard figures, the audio frequency spectrum is often divided into three regions: bass, midrange, and treble. The low frequencies (bass) are those between 20 and 250 hertz and provide the "power," or "bottom," to sound. Too little bass gives a "thin" sound, and too much bass gives a "boomy" sound. The midrange frequencies fall between 250 and 4500 hertz. These frequencies provide a lot of sound's substance and intelligibility. Too little midrange gives a lack of "presence," but too much midrange gives a "harsh" sound. High frequencies (treble) are those from 4500 hertz to 20,000 hertz. The treble frequencies provide the brilliance and sharpness to sound. Too little treble gives a "dull" sound, and too much treble gives an excess "sparkle" to the sound as well as increasing the likelihood of hearing noise or hiss in the sound.

As frequencies change, we think in terms of the musical interval of the **octave**, or a change in pitch caused by doubling or halving the original frequency. For example, a sound going from bass to midrange to treble frequencies by octave intervals would go from 110 hertz to 220 hertz to 440 hertz to 880 hertz to 1760 hertz to 3520 hertz to 7040 hertz and so on. As humans, we are subject to an awkwardly named **equal loudness principle**, by which we hear midrange frequencies better than either high or low frequencies. In radio production (and other forms of sound manipulation), we often compensate for this by equalization of the signal.

2.14 EQUALIZERS AND PAN POTS

Many boards have simple equalizers (EQ). These increase or attenuate certain frequencies, thus altering the sound of the voice or music by changing the tonal quality of the

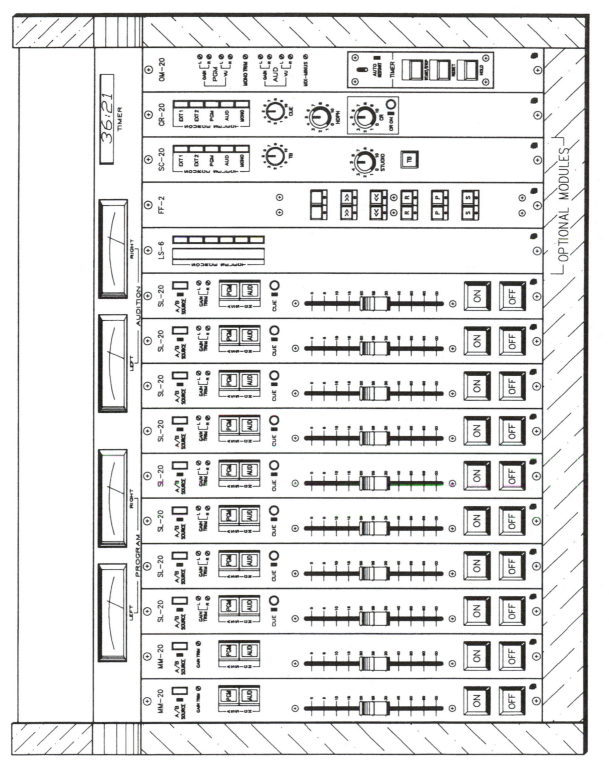

FIGURE 2.8 Audio console with numerous features. (Courtesy of Wheatstone Corporation)

sound. In some instances they eliminate unwanted sound. For example, scratches in records are heard mainly on high frequencies. By filtering out these frequencies, the record will sound scratch free. Likewise, a low rumble can be removed by eliminating or turning down low frequencies. Equalizers can also be used for special effects, such as making a voice sound like it is coming over a telephone line. It's important to note that when you equalize a sound, you affect both the unwanted and wanted sound; equalization is usually a compromise between eliminating a problem and keeping a high-quality, usable audio signal.

Usually the equalizers are knobs or switches that increase or attenuate a certain range of frequencies. They are placed somewhere above each input volume control. As shown in Figure 2.5, the input signal of this channel is split into three frequency ranges: high, midrange, and low. Turning the control clockwise increases the volume of that range of frequencies; counterclockwise rotation of the control will decrease its level. The "EQ in" button on this channel is actually a bypass switch that allows the operator to hear and compare the signal with or without equalization. Equalizers and other signal processing equipment will be discussed in further detail in Chapter Nine. Audio console channel inputs that are **monaural** (such as a microphone channel) often have a **pan pot**, or **pan knob**. By turning (panning) this knob to the left or right, you can control how much of the sound from that input goes to the right channel and how much goes to the left channel output. In other words, if the pan pot of a mic channel is turned toward the *L* position, upon monitoring, the vocal would sound stronger, or louder, from the left speaker. Normally, the pan pot would be in the center position and the input sound would be directed equally to the left and right outputs of that channel. **Stereo** input channels may have a **balance control**, which serves a similar purpose.

2.15 TONE GENERATORS

Some boards have a built-in **tone generator**. This reference tone is usually placed on a tape before the program material. The tone generator sends out a tone through the board that can be set at 100 percent, using the board VU meter. The VU meter on the source to which the signal is being sent (e.g., an audio tape recorder) is simultaneously set at 100 percent. After the two are set, any other volume sent

through the board will be the exact same volume when it reaches the tape recorder. Having a tone generator allows for this consistency. Otherwise, sounds that register at 100 percent coming through the board might peak in the red and be distorted on the tape recorder.

The tone on the tape is also used when the tape is played back. The audio engineer or board operator listens to the tone and sets that tape recorder VU meter to 100 percent. That way the tape will play back exactly as it's recorded.

2.16 OTHER FEATURES

Some audio boards have a **solo switch** above each input. When this switch is on, only the sound of that particular input will be heard over the monitor. Some boards have a **talk-back switch**, which is a simple intercom system consisting of a built-in microphone and a pushbutton control that turns the mic on or off. The normal position of this switch is off so that the button must be pushed in to activate the mic. The signal from the talk-back mic is sent to a speaker in another studio—for example, a performance studio—which would allow the operator at the audio console to communicate with the announcer in a studio at a separate location.

A **mute switch** on an audio console channel prevents the signal from going through that channel when it's depressed. A mute switch acts just like an on/off switch.

Boards also have provisions for echo, reverberation, and some interface with computers. These functions will be discussed in Chapters Nine and Ten. To see if you understand the various functions of an audio console, look at Figure 2.8 and determine if you can tell what each control does.

2.17 SOUND TRANSITIONS AND ENDINGS

As mentioned early in this chapter, one of the functions of the audio console is to mix two or more sound sources together. Often this mix is really a sound transition, or the merging of one sound into another. In radio production, the basic transition is the **fade** (gradually increasing or decreasing volume), where you mix one sound with silence. For example, to **fade-in** a CD means to slowly increase the volume from silence to the desired level (see Figure 2.9A). A **fade-out** accomplishes just the opposite as the CD goes

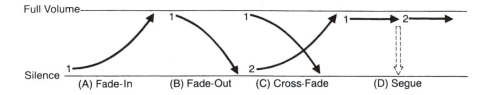

FIGURE 2.9   Sound transitions a) Fade-in; b) Fade-out; c) Cross-fade; d) Segue.

from normal, full volume to silence (see Figure 2.9B). Most compact discs are recorded so that they fade out at the end naturally, but in production work it's sometimes necessary to end a song early, and the production person can do so with a manual fade-out. The other common transitions are the cross-fade and the segue. As the name implies, a **cross-fade** occurs when one sound is faded down as another sound is faded up (see Figure 2.9C). There is a point as the two sounds cross when both sounds are heard. Because of this, care should be taken in choosing music to cross-fade. Some combinations of songs can sound extremely awkward. The speed of a cross-fade is determined by the board operator and depends on the type of effect desired; however, most cross-fades are at a medium speed to give a natural, brief blending of the two sounds. A **segue** is quite different; it is the transition from one sound to the next with no overlap or gap (see Figure 2.9D). A segue can best be accomplished when the first song ends cold. Music with a **cold ending** doesn't fade out, but rather it has a natural, full-volume end. Unlike fades of any kind, the segue is accomplished with both sounds at full volume. Most disc jockey work is a mixture of cross-fades and segues, but all the sound transitions are used frequently in radio production work.

In addition to the fade-out and cold ending, one other music ending that should be noted is the **sustain ending**. Songs with a sustain ending will hold the final chord or notes for a short period of time and then very gradually fade out. This is distinctly different from either a normal fade-out or a cold ending. Music that is specifically designed for radio use will often have an indication of the ending—F, C, or S—as part of the timing information about the song.

2.18 CONCLUSIONS

If you have followed the descriptions and explanations offered in this chapter, the audio console should be a less frightening assemblage of switches, knobs, and meters than it was when you began. You should begin to have a good idea of how to operate each board and feel comfortable working with the controls of your own board. You should also be aware that the audio console of the future may be a virtual audio console, and you'll manipulate the controls with a few mouse clicks.

Self-Study

■ Questions

1. It's possible to have music from an audio tape recorder go into a control board and then come out and be recorded on another tape recorder.
   a) true
   b) false

2. In Figure 2.5, according to the pan pot position, what is the relationship between the sound signal going to the left channel and the signal going to the right channel?
   a) left would be less than right
   b) left would be the same as right
   c) left would be greater than right
   d) there would be no signal going to the right

3. In Figure 2.3A, if the fader were at −5 and you changed it to −25, you would have _____.
   a) amplified the signal
   b) put the channel in cue
   c) decreased the resistance
   d) decreased the volume

4. In Figure 2.3A, sound would not get to line out on channel 1 because _____.
   a) the fader is at 0
   b) the output selector switch is off
   c) the program amplifier is off
   d) the input selector switch is set to mic

5. In Figure 2.4, audible sound comes from the _____.
   a) VU meter
   b) monitor amplifier
   c) program amplifier
   d) monitor speaker

6. In Figure 2.6, 50 percent on the scale is roughly equivalent to _____.
   a) –6 dB
   b) –4 dB
   c) 50 dB
   d) –10 dB

7. The relative position of the needle of the VU meter below is an example of _____.
   a) peaking in the red
   b) turning up the pot
   c) riding in the mud
   d) riding the gain

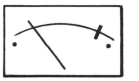

8. The relative position of the needle of the VU meter below is an example of _____.
   a) riding the gain
   b) broadcasting in stereo
   c) peaking in the red
   d) pegging the meter

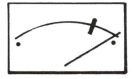

9. The line position on an input selector would be used to bring in _____.
   a) CDs and audio tape recorders
   b) microphones and turntables
   c) microphones and tape recorders
   d) only microphones

10. The monitor/speaker is an example of _____.
    a) an input
    b) a mix
    c) an output
    d) an equalizer

11. If the program/audition switch is in the audition position _____.
    a) sound will not reach the mic/line switch
    b) sound can be going to a tape recorder
    c) sound can be going to the transmitter
    d) sound will go to the cue speaker

12. The master volume control _____.
    a) must be up for sound to leave the board
    b) is required only if the board is stereo
    c) controls only the volume of the line inputs
    d) controls only the volume of the mic inputs

13. Which control could be used to help eliminate scratches on records?
    a) the pan knob
    b) the gain trim
    c) the solo button
    d) the equalizer

14. The cue position on a fader _____.
  a) allows sound to go to the transmitter
  b) sometimes substitutes for the trim control
  c) sends sound to a small speaker in the audio board
  d) allows sound to fade from left to right channel

15. Which of the following can help assure that the level that's being recorded on a tape recorder is the same as that coming from the audio board?
  a) tone generator
  b) remote switch
  c) digital timer
  d) pan pot

16. In Figure 2.8, if channels 1 and 6 were in audition position, channels 2, 3, 4, and 5 were in program position, and channel 3 was also in cue, then sound would get to the transmitter from _____.
  a) inputs 1, 2, and 3
  b) inputs 2, 4, and 5
  c) inputs 4, 5, and 6
  d) all six inputs

17. When a disc jockey fades down one CD (or record) at the same time as another CD (or record) is faded up, this is known as a _____.
  a) fade-in
  b) fade-out
  c) segue
  d) cross-fade

18. On an audio console, the term _____ refers to a group of switches, faders, and knobs that are usually associated with one sound source, such as a CD player.
  a) input selector
  b) remote start switch
  c) channel
  d) output selector

19. A segue is the basic sound transition in which one sound is "mixed" with silence.
  a) true
  b) false

20. Most audio consoles used in radio broadcasting are identical to the consoles used in music recording.
  a) true
  b) false

21. Which feature of an audio console would allow the operator to alter the tonal quality of a sound going through the board?
  a) mute switch
  b) gain trim control
  c) equalizer control
  d) talk-back switch

22. A channel of the audio console shown in Figure 2.8 can be put into cue by _____.
  a) moving the fader up to "0"
  b) moving the fader down to "60"
  c) depressing the "on" button
  d) depressing the "cue" button

23. Which feature of an audio console is a simple intercom system?
    a) talk-back switch
    b) tone generator
    c) pan knob
    d) output selector

24. Because of the complexity of the numerous buttons, switches, and knobs associated with the audio console, this is one piece of production equipment that is *not* likely to be moving from an analog to a digital configuration.
    a) true
    b) false

25. As a review of audio boards, match the items in the top list (1, 2, 3 . . .) with the choices in the bottom list (a, d, m . . .), and then select the correct set of answers from the sequences shown in a, b, c, or d below.

    1. _____ mic/line switch
    2. _____ fader
    3. _____ VU meter
    4. _____ program/audition switch
    5. _____ preamp
    6. _____ monitor amp
    7. _____ program amp
    8. _____ pan pot
    9. _____ equalizer
    10. _____ talk-back switch

    a. amplifies sound before it goes to the pot
    d. determines how much preamplification a signal will receive
    m. enables a mic built into the board to be used to talk to another studio
    t. determines whether sound goes to the transmitter or stays within the control room
    s. amplifies sound before it goes to a monitor
    l. amplifies sound before it goes to line out
    v. varies resistance
    r. peaks in the red
    f. attenuates or boosts certain frequencies
    c. controls the amount of sound going to left and right channels

    a) 1.d 2.v 3.r 4.t 5.a 6.s 7.l 8.c 9.f 10.m
    b) 1.t 2.m 3.r 4.d 5.s 6.l 7.a 8.c 9.f 10.v
    c) 1.d 2.v 3.r 4.t 5.s 6.l 7.a 8.c 9.f 10.m
    d) 1.d 2.m 3.r 4.t 5.a 6.s 7.l 8.f 9.c 10.v

■ ANSWERS

If You Answered A:

1a. Right. A tape recorder can be both input and output.
2a. No. Check the setting of the pan pot. (Reread 2.14.)
3a. No. The fader never amplifies the signal. This is done at the preamp. (Reread 2.5.)
4a. Wrong. When a pot is at 0, it is on. (Reread 2.5 and 2.10.)
5a. No. This indicates level, but you hear nothing from it. (Reread 2.6 and 2.9)
6a. Right. You read the scale correctly.
7a. No. The needle would be at the other end for this. (Reread 2.9)
8a. No. Riding gain means to keep the volume at proper levels, and this is not being done. (Reread 2.9.)
9a. You're correct. Both of these are inputs that are already amplified.
10a. No. You're at the wrong end. (Reread 2.2.)
11a. No. The mic/line switch is long before the program/audition switch and really has nothing to do with it. (Reread 2.4 and 2.10.)

12a. Right. The purpose of the master volume is to allow all appropriate sounds to leave the audio console.

13a. No. This controls how much sound is going to the left and right channel. (Reread 2.14.)

14a. No. It specifically prohibits it from going to the transmitter. (Reread 2.7.)

15a. Right. If the board VU meter and the tape recorder VU meter are both set at 100 percent tone, the levels should be the same.

16a. No. It wouldn't get there from either input 1 or input 3. For input 1, the program/audition switch is in the audition position. For input 3, the fader is in the cue position. (Reread at least 2.7 and 2.10, but if you had a great deal of difficulty figuring this out, reread the whole chapter.)

17a. No. This is another sound transition. (Reread 2.17.)

18a. No. This is one of the switches involved, but this is not the correct term. (Reread 2.3 and 2.4)

19a. No. You've confused this with the fade. (Reread 2.17.)

20a. No. While there are similarities between all audio consoles, those most often found in radio are more basic than those used in the music recording studio. (Reread 2.1.)

21a. Wrong. This control turns off a console channel. (Reread 2.16.)

22a. No. This will turn the volume of the channel up, probably to a distorted level. (Reread 2.5 and 2.7.)

23a. Correct. The talk-back feature allows a board operator to talk with an announcer in another studio via a simple intercom system.

24a. No. You're wrong. Digital, virtual consoles are already a part of production-room equipment. (Reread 2.1 and 2.18.)

25a. Right. You chose the correct sequence.

If You Answered B:

1b. Wrong. A tape recorder can be fed in and another tape recorder can be placed at the output. (Reread 2.2 carefully.)

2b. No. Check the setting of the pan pot. (Reread 2.14.)

3b. No. Cue most likely would be even below the −55 mark. (Reread 2.5 and 2.7.)

4b. Right. The output selector switch is in the off position, so sound would stop there.

5b. No. This amplifies so that the sound can come out, but you don't hear the sound from it. (Reread 2.6.)

6b. No. You went the wrong direction. (Reread 2.9.)

7b. No. If anything, the pot is being turned down. (Reread 2.9.)

8b. No. (Reread 2.9.)

9b. No. A microphone should come in a mic input because it needs to be amplified to reach the level of amplification of a turntable. The turntable would be plugged into the line-in position. (Reread 2.4.)

10b. No. This is way off. (Reread 2.2.)

11b. Yes. The purpose of the audition position is to send the sound to someplace other than the transmitter. Sound would not necessarily need to go to a tape recorder, though it could.

12b. No. A master volume control functions the same for both mono and stereo. Stereo boards, however, usually have two master volume controls, one for each channel. (Reread 2.11.)

13b. Wrong. This allows finetuning of input or output levels on an audio board. (Reread 2.5 and 2.14.)

14b. No. The two have nothing to do with each other. (Reread 2.5 and 2.7.)

15b. No. This will turn on the recorder remotely but will do nothing about levels. (Reread 2.12 and 2.15.)

16b. Right. For all of these channels the volume is up, the switch is in program, and the master volume controls are up.

17b. No. This is another sound transition. (Reread 2.17.)

18b. No. This is one of the switches involved, but this is not the correct term. (Reread 2.3 and 2.12.)

19b. Correct. The basic sound transition that mixes one sound with silence is the fade, not the segue.

20b. Yes. This is the correct response.

21b. Wrong. This control varies the input level of a sound source of a console channel. (Reread 2.4 and 2.5.)

22b. No. While some consoles are put into cue by moving the fader down, it would have to go all the way down and usually "click" into a cue position. (Reread 2.5 and 2.7.)

23b. Wrong. The tone generator provides a reference level for setting correct recording levels. (Reread 2.15.)

24b. Yes. This is a correct response because both digital and virtual audio consoles are readily available now.

25b. No. You made many errors. (Reread the chapter.)

If You Answered C:

2c. Right. The pan pot is set so that more of the sound from the input goes to the left channel.

3c. You're warm but not right. Changing the level from −5 to −25 would increase resistance. (Reread 2.5.)

4c. Wrong. The program amp, like the preamp, is always on. (Reread 2.5 and 2.10.)

5c. No. This amplifies sound but you don't hear sound from it. (Reread 2.6.)

6c. There is no such thing on a VU meter. (Reread 2.9, review Figure 2.6.)

7c. Right. "Riding in the mud" is the term for a low reading.

8c. No. It is worse than that. (Reread 2.9.)

9c. No. A microphone should come in a mic input because it needs to be amplified to reach the level of amplification of a tape recorder. The tape recorder would be plugged into the line in position. (Reread 2.4.)

10c. Right. Sound comes out to the monitor/speaker.

11c. No. The purpose of the audition position is to keep the sound from going to the transmitter. (Reread 2.10.)

12c. No. It controls all the sound that is set to leave the board. Line and mic positions have no bearing on it. (Reread 2.4 and 2.11.)

13c. No. This allows you to hear one input by itself. (Reread 2.14 and 2.16.)

14c. Right. Cueing is just for the person operating the board.

15c. No. This is simply a clock. It has nothing to do with levels. (Reread 2.12 and 2.15.)

16c. No. The sound wouldn't get there from input 6 because the program/audition switch is in audition. (Reread 2.10.)

17c. No. This is another sound transition. (Reread 2.17.)

18c. Yes. This is the correct term.

21c. Right. EQ controls increase or decrease certain frequencies of the sound thus changing the tone.

22c. No. If the fader was up, this would let us hear audio, but not in cue. (Reread 2.12.)

23c. Wrong. Pan pots control the amount of sound that goes to the left or right output of a channel. (Reread 2.14.)

25c. No. You're confused about the various types of amplification. (Reread 2.4, 2.6, 2.9, and 2.10.)

If You Answered D:

2d. No. That would not be correct. (Reread 2.14.)

3d. Right. This is an increase in resistance to the signal; not as much of the signal gets through, and it's softer.

4d. Wrong. It wouldn't matter if the input was set to mic or line. (Reread 2.4 and 2.10.)

5d. Right. You hear sound from the monitor speaker.

6d. You might be thinking half of the dB scale, but that isn't correct. (Reread 2.9.)

7d. No. If anything, the operator is not riding the gain. (Reread 2.9.)

8d. Right. Pegging the meter is correct. We could also have used the term "pinning the needle."

9d. Wrong. Microphone signals must be amplified to reach line level and should only come in a mic input. (Reread 2.4.)

10d. No. An equalizer affects frequencies. (Reread 2.2 and 2.14.)

11d. Wrong. The cue speaker will be activated when a channel is put into cue, regardless of where the program/audition switch is. (Reread 2.7 and 2.10.)

12d. No. It controls all the sound that is set to leave the board. Mic and line positions have no bearing on it. (Reread 2.4 and 2.11.)

13d. Right. This can eliminate high frequencies where scratches reside.

14d. No. You're confusing this with a pan pot. (Reread 2.7 and 2.14.)

15d. No. (Reread 2.14 and 2.15.)

16d. Wrong. You probably don't really understand the functions of program/audition switches and cue. (Reread 2.1, 2.3, 2.7, and 2.10.)

17d. This is the correct response.

18d. No. This is one of the switches involved, but this is not the correct term. (Reread 2.3 and 2.10.)

21d. Wrong. This control is a form of studio intercom. (Reread 2.16.)

22d. Yes. On this console, any channel can be put into cue merely by pressing the cue button associated with that channel, regardless of where other switches or buttons are set.

23d. Wrong. Output selectors determine where sound goes when it leaves the audio console. (Reread 2.10.)

25d. No. You have confused equalizer and pan pot. (Reread 2.14.)

Projects

■ PROJECT 1

Learn to operate an audio console.

Purpose

To acquaint you with the operation of an audio console, to make you somewhat proficient at some of its functions, and to practice the basic sound transitions.

Advice, Cautions, and Background

1. Audio boards are generally the most complicated pieces of equipment in a radio station. It may take you a while to master a board, but don't despair. Take it slowly, and don't be afraid to ask for help.
2. Audio boards all have the same general purpose. Sounds come into the board, are mixed together, and are sent out to somewhere else.
3. The actual exercise should be done as quickly as possible. You won't be judged on aesthetics. In other words, when you're fading from one source to another, do it quickly. Don't wait for the proper musical beat, phrase, or pause.

How to Do the Project

1. Familiarize yourself with the operation of the audio console in your production studio. Learn the inputs, the outputs, the method for changing volume, and other special features of the board.
2. As soon as you feel you understand the board, do the following exercise as rapidly as possible while recording it on a tape recorder. Practice as much as you like first.
   a. Cue up a CD, play part of it, and fade it out.
   b. Using the studio mic, announce your name and the time.
   c. Begin a second CD.
   d. Cross-fade to another CD, and then fade it out.
   e. Bring in an auxiliary microphone, and ad-lib with another announcer
   f. Fade in either a record or a CD, segue to another CD, fade it under.
   g. Announce something clever on the studio mic.
3. Listen to your tape to make sure it recorded properly.
4. Label your tape AUDIO CONSOLE OPERATION, and hand it in to your instructor for credit for the project.

■ PROJECT 2

Diagram and label the audio console in your production room.

Purpose

To familiarize you with the positioning of the various switches and controls so that you can access them quickly.

Advice, Cautions, and Background

1. Some boards are very complicated and have more functions than are discussed in this chapter. Usually this is because they're intended to be used for sound recording of music. If you have such a board, you only need to label the parts that you will be using frequently.
2. If you can't find controls for all of the functions given in this chapter, ask for help. Because there are so many different brands and types of boards, sometimes functions are combined or located in places where you cannot identify them easily.

3. You don't need to label each switch and knob. If your board has ten inputs, it will obviously have ten channel volume controls. You can circle them all and label them together, or make one label that says INPUT VOLUME CONTROLS, and draw arrows to all ten.
4. You will be judged on the completeness and accuracy of your drawing. You won't be graded on artistic ability, but be as clear as possible.

How to Do the Project

1. Sketch the audio console in your production studio.
2. Label all the basic parts: input selectors, channel gain controls, VU meters, output selectors (program/audition switches), and master gain controls.
3. Also label any other parts of the board that you will be using frequently, such as equalizers, cue positions, and headphone connections.
4. If possible, give the brand name and model number of the board.
5. Label your sketch AUDIO CONSOLE, and give your completed drawing to the instructor for credit for this project.

CHAPTER THREE

Microphones

Information

3.1 INTRODUCTION

In the radio production room, the microphone takes on an important role. It's the piece of equipment that changes the announcer's voice into an electrical signal that can then be mixed with other sound sources and sent to a recorder or broadcast over the air. Because the purpose of the microphone is to change sound energy into electrical energy, it's called a **transducer** (a device that converts one form of energy into another).

3.2 CLASSIFYING MICROPHONES

There is no one correct mic to use in radio production work, but certain types of microphones will work better than others in certain situations. Microphone types are usually determined by three factors: the sound-generating element of the mic, the pickup pattern of the mic, and the impedance of the mic.

Categorized by their sound-generating element, there are two types of microphones commonly used in radio: the dynamic mic and the condenser mic.

3.3 DYNAMIC MICROPHONES

The **dynamic mic** is sometimes known as the **moving-coil microphone**, or to a lesser extent, the **pressure mic**. This microphone's sound-generating element is constructed of a diaphragm, a permanent magnet, and some coils of wire wrapped around the magnet. The diaphragm is positioned within the field of the magnet and responds to the pressure of the sound. Movements of the diaphragm caused by sound waves result in a disturbance of the magnetic field, and this induces a small electric current into the coils of wire (see Figure 3.1). In a later chapter, you'll see that the basic loudspeaker consists of similar elements and works like a dynamic microphone in reverse by changing electrical energy into sound energy.

The dynamic mic is very commonly used in radio and has many advantages that make it so popular. It's a small, fairly inexpensive mic, yet it has excellent **frequency re-**sponse so that both highs and lows reproduce accurately. The reason it has seen such acceptance by broadcasters is its sturdy design.

The dynamic mic can withstand a moderate amount of abuse (which often occurs in the broadcast setting). It's also fairly insensitive to wind, and this, along with its ruggedness, makes it an excellent remote mic. The dynamic mic can be used in most broadcast situations—as a stand mic, a handheld mic, or a lavalier (a small mic hung around the neck or clipped to clothing below the neck). The main disadvantage of the dynamic mic is that it does not satisfactorily reproduce the voices of certain individuals.

With some announcers, the mic exaggerates plosives (popping on *p*'s) and sibilance (hissing on *s*'s).

Even though the dynamic mic is fairly rugged, all microphones are fragile to some extent and should be handled with care like any other piece of audio production equipment. Beginning announcers often misuse a mic by blowing into it to see if it's live or to set a level. This is the worst way to test a mic and can result in serious damage to the microphone. In fact, the higher the quality of the microphone, the more likely that it *will* be damaged in this manner.

3.4 CONDENSER MICROPHONES

The other common type of broadcast microphone is the **condenser mic**. Also known as a **capacitor mic**, it uses an electronic component—the capacitor—to respond to the sound. The sound-generating element consists of a charged conductive diaphragm and metallic backplate separated by an insulating material. The diaphragm responds to sound waves, changing the distance between the diaphragm and the backplate; this alteration changes the capacitance and generates a small electrical signal (see Figure 3.2).

The condenser mic requires a power supply, such as a battery, to charge the backplate and diaphragm and related electronic components. Because batteries used to be large and cumbersome, early condenser mics were both inconvenient and expensive. Today's condenser mics utilize small internal power supplies or **phantom power** supplies. The latter usually comes from a recorder or an audio board through the mic cable to the microphone. The condenser

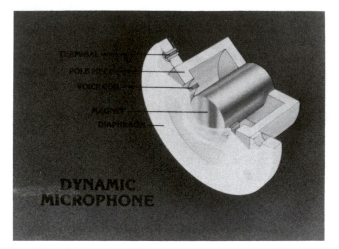

FIGURE 3.1 Internal structure of a dynamic microphone. (Courtesy of Shure Brothers, Inc.)

microphone is an excellent radio mic because it's fairly rugged and produces excellent sound quality and wide frequency response. While the dynamic mic is the most-used radio production mic, the condenser mic is also frequently found in the modern radio production studio. In addition, the built-in microphone on modern portable cassette recorders is usually a condenser mic that provides fairly good quality on both consumer and professional models.

Another form of condenser microphone is the **electret mic**. Its capacitor is permanently charged during manufacture, but it still needs a power supply (albeit much less power) for its electronic circuit. The lower power requirement offers some flexibility in usage, but the electret mic possesses less dynamic range than the regular condenser microphone, so you may not find an electret mic in many radio production studios.

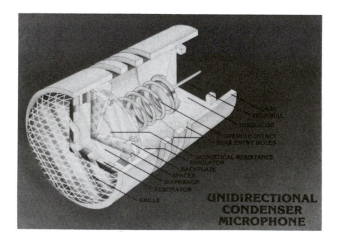

FIGURE 3.2 Internal structure of a condenser microphone. (Courtesy of Shure Brothers, Inc.)

3.5 OTHER MICROPHONES

There are other types of microphones. For many years the **ribbon mic** was common in radio. Another form of a dynamic microphone, the ribbon mic contained a sound-generating element that consisted of a thin metallic ribbon stretched in the field of a magnet. Sound waves vibrating the ribbon generated an electrical signal.

The ribbon mic had an excellent warm, smooth sound, but it was bulky and fragile and has been largely replaced by the condenser mic, which has a similar high-quality sound.

A **regulated phase microphone** can be thought of as part dynamic mic and part ribbon mic. A wire coil is attached to or impressed into the surface of a circular diaphragm. The diaphragm is suspended between two circular magnets that are designed to be acoustically transparent. An electrical current is generated as sound waves vibrate the diaphragm. Because of its design, the regulated phase mic exhibits some of the quality of a ribbon mic with some

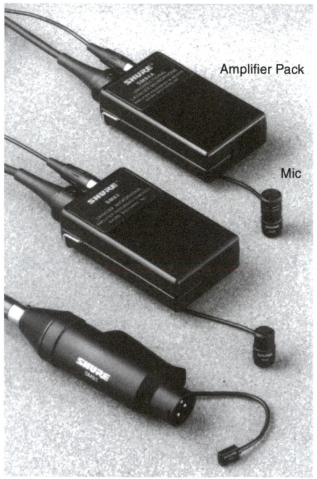

FIGURE 3.3 Lavalier microphones. (Courtesy of Shure Brothers, Inc.)

of the ruggedness of the dynamic mic. This mic is used more in the recording industry than in radio, but you may find one in some production studios.

3.6 SPECIAL PURPOSE MICROPHONES

There are other microphone "types," but these are not based on internal sound-generating elements. In fact, these microphones usually employ one of the basic designs mentioned in the previous sections; however, they are often designed for a specific usage and are frequently considered another variety of mic. For example, the **lavalier mic** (see Figure 3.3) mentioned previously is a tiny microphone, usually less than 1 inch long, that can be unobtrusively clipped to an announcer's lapel or tie. There are both dynamic and condenser models, and while the lavalier mic is occasionally used in a radio remote situation, its small size makes it more appropriate for television than radio.

A **stereo microphone** like the one shown in Figure 3.4 incorporates small, multiple sound-generating elements within a single mic housing that can duplicate the various stereo miking techniques. You'll learn more about these stereo miking techniques later in this chapter.

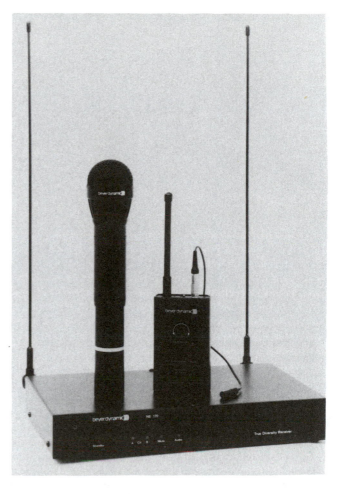

FIGURE 3.5 FM wireless microphone. (Courtesy of Beyerdynamic)

FIGURE 3.4 Stereo microphone—designed for X-Y miking. (Courtesy of AKG Acoustics, Inc.)

FM microphones—also known as **RF** (radio frequency), **wireless**, and **radio mics**—are used occasionally in radio production situations when a microphone cable might hinder the recording or production. An FM mic is really a miniature radio station (see Figure 3.5). The audio signal generated by the mic (employing either a dynamic or condenser element) is sent out from the mic by a low-power transmitter rather than a microphone cable. This transmitter is either in the mic housing or contained in a small pack designed to be worn by the announcer. The transmitted signal is picked up by a receiver located nearby and converted from a radio frequency signal back into an audio signal at that point. One problem associated with wireless mics is interference. Since they broadcast on specific FCC-assigned frequencies that are not exclusive to FM mics, they sometimes pick up interference from VHF TV, cordless telephones, and other radio frequency users.

The **pressure zone mic**, or the **PZM** (also known as a **boundary**, **plate**, or **surface-mount mic**), is a small microphone capsule mounted next to a sound-reflecting plate (or

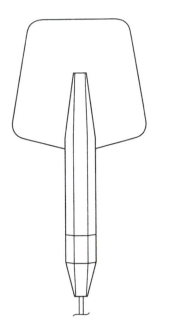

FIGURE 3.6 PZM microphone design. (Courtesy of Crown International, Inc.)

boundary) as shown in Figure 3.6. Designed to be used on a flat surface, such as a tabletop, the mic picks up sound from all directions above the table surface. Boundary mics offer exceptional microphone sensitivity because both direct sound and reflected sound reach the mic at the same time since the mic is so close to the reflective surface. This boosts the incoming sound signal, which improves the clarity of the microphone.

Figure 3.7 shows that the **shotgun microphone** is appropriately named. A mic capsule is at one end of a tube, or barrel, that is "aimed" like a gun toward the sound source. The design of the mic rejects sound from the side and rear but picks up a very narrow angle of sound from the front of the mic's barrel. The highly directional nature of the mic makes it good at picking up sound from a considerable distance from the sound source; however, sound quality is somewhat less than mics that pick up nearby sounds.

There are other specialized microphones available, but in typical radio production work you aren't likely to run across them. In fact, even the mics mentioned in this section will only be useful on occasion, but an understanding of them may prove helpful from time to time.

3.7 MICROPHONE PICKUP PATTERNS

Another way of classifying microphones is by their pickup patterns. Microphones can be constructed so that they have different directional characteristics. In other words, they pick up sound from varying directions as shown in Figure 3.8. Sound picked up at the front of the microphone, or at 0 degrees, is said to be on-axis. Sound picked up from the mic's side is 90 degrees off-axis, and sound picked up from the rear of the microphone is 180 degrees off-axis. The microphone housing, by the use of small openings and ports, is designed so that unwanted sound (often off-axis sound) is canceled out or attenuated as it enters the mic. The on-axis sound received directly from the front of the mic fully impacts the mic's diaphragm.

Three basic sound patterns are omnidirectional, bidirectional, and cardioid. While most microphones will have one fixed pickup pattern, **multidirectional mics** have switchable internal elements that allow the microphone to employ more than one pickup pattern.

3.8 THE OMNIDIRECTIONAL PICKUP PATTERN

The **omnidirectional** mic is also known as a **nondirectional** mic. These two terms may seem to contradict each other—*omni* (all) and *non* (no). Both terms are correct, however. The mic picks up sound in all directions, but it also has no particular pickup pattern. Omnidirectional mics pick up sound equally well in just about any direction. Think of an orange with a microphone right in the middle. No matter where the sound comes from, the mic (the orange) responds to it equally well. Figure 3.9 illustrates the pickup pattern for a typical omnidirectional microphone.

Omnidirectional mics are used whenever it is desirable to pick up sound evenly from all sides of the mic, including

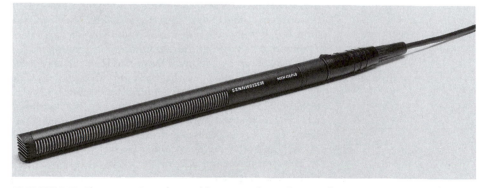

FIGURE 3.7 Shotgun microphone. (Courtesy of Sennheiser Electronic Corporation)

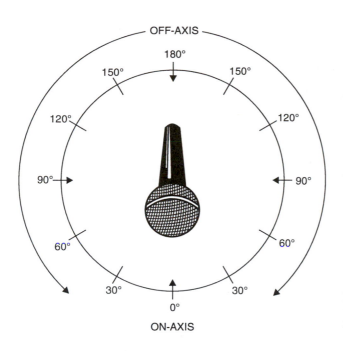

FIGURE 3.8 Microphone's axis.

above and below it. Omnidirectional mics are commonly used in broadcast situations outside the studio when the ambience of the location needs to be picked up along with the announcer's voice.

Of course, the fact that they pick up sound equally well from all directions can also be a disadvantage. You may pick up unwanted background noise (such as traffic noise) in addition to the announcer when recording in a remote situation. Omnidirectional mics used in a highly reflective room may also produce a "hollow" sound, since they tend to pick up more reverb than other types of mics.

3.9 THE BIDIRECTIONAL PICKUP PATTERN

The **bidirectional** mic picks up sound mainly from two directions: the front and the rear of the mic, or on-axis and off-axis (see Figure 3.10). Its pickup pattern can be visualized as a figure 8, with the microphone located at the intersection of the two circles, or, to maintain the fruit analogy, as two grapes side-by-side. It was often used for radio dramas so that actors could face each other, but it is not a common pickup pattern for today's broadcast mics. Although you may not utilize it in radio production situations, it's a good mic for the basic two-person interview.

3.10 THE CARDIOID PICKUP PATTERN

The **cardioid** mic is sometimes referred to as **unidirectional** because it picks up sound mainly from one direction, the front of the microphone. Its pattern is actually heart-

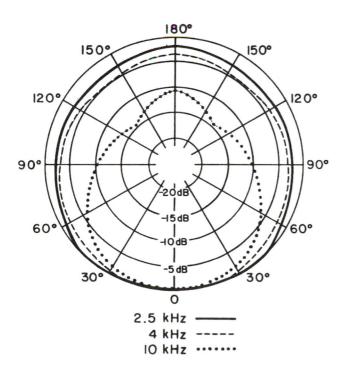

2.5 kHz ———
4 kHz -----
10 kHz ••••••

FIGURE 3.9 Omnidirectional polar response pattern. (Courtesy of Shure Brothers, Inc.)

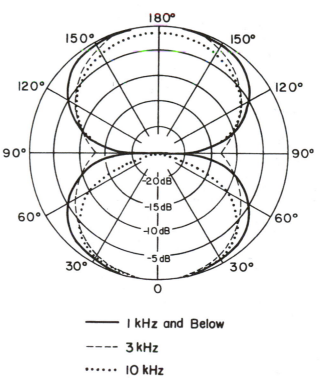

——— I kHz and Below
----- 3 kHz
•••••• 10 kHz

FIGURE 3.10 Bidirectional polar response pattern. (Courtesy of Shure Brother's, Inc.)

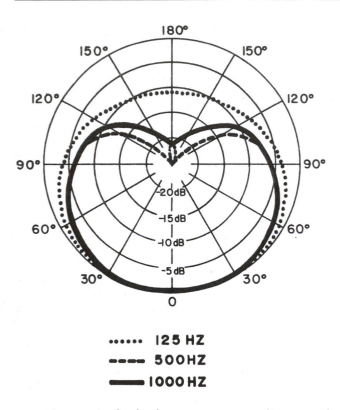

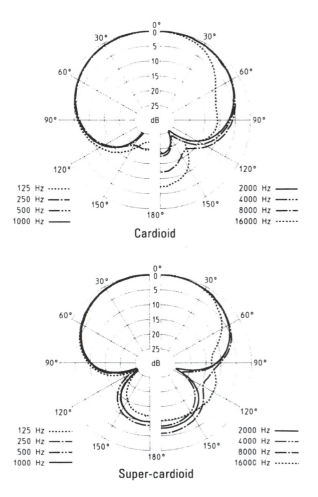

FIGURE 3.11 Cardioid polar response pattern. (Courtesy of Shure Brothers, Inc.)

the difference between **pickup pattern** and **polar pattern**. We have already described a microphone's pickup pattern as the area around the microphone in which the mic best "hears" or picks up the sound. This is a three-dimensional shape, as demonstrated by the orange, grape, and apple analogies. The diagrams shown, however, are two dimensional. When a microphone's pickup pattern is shown by a two-dimensional drawing, we call that drawing the mic's polar pattern, or polar response pattern. Compare the cardioid pickup pattern shown in Figure 3.13 with the cardioid polar response pattern shown in Figure 3.11.

A polar pattern is drawn around a full circle, or 360 degrees. As noted earlier, the line from 0 degrees to 180 degrees represents the mic's axis, and sound entering the mic from the front (0 degrees) is on-axis. The concentric circles show decreasing dB levels and allow us to see how the sound will be attenuated as it is picked up off-axis. For example, Figure 3.11 shows that sound picked up directly in front of the mic will experience no attenuation. On the other hand, sound picked up from either side (90 degrees off-axis) will be attenuated about 7 dB, and sound coming di-

shaped—hence the name cardioid (refer to Figure 3.11). Another way to visualize its pickup pattern is to think of an upside-down apple, of which the stem represents the mic and the rest of the apple approximates the cardioid pickup pattern. Although the mic picks up sound from the front and sides, the level of pickup from the sides is only about half that of the front. It doesn't pick up sound well from the rear at all (less than one tenth of the sound from the front).

Three variations on the cardioid are the **supercardioid**, **hypercardioid**, and **ultracardioid** microphones. They continue to offer great rejection of the sound from the sides, but each mic also picks up a narrower scope of on-axis sound (see Figure 3.12). In other words, each of these mics becomes more and more directional in nature.

All cardioid mics are popular because they do reject unwanted sounds (excessive reverb, feedback, background noise), but the announcer must be careful to stay "on mic" and not move too far off-axis, especially when using super-, hyper-, or ultracardioid mics. Cardioid mics are often employed in sports remote broadcasts; the sportscaster talks into the live side (front) of the mic, and the crowd and action sounds coming from the rear and sides are limited.

3.11 POLAR RESPONSE PATTERNS

The diagrams shown in Figures 3.9–3.12 indicate polar patterns. One of the concepts that you need to understand is

FIGURE 3.12 Polar response pattern of cardioid and supercardioid. (Courtesy of Sennheiser Electronic Corporation)

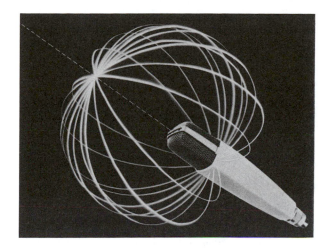

FIGURE 3.13 Three-dimensional view of cardioid pickup pattern. (Courtesy of Sennheiser Electronics Corporation)

rectly from the rear of the mic (180 degrees) will be attenuated about 20 dB. The solid, dashed, and dotted lines show that microphone patterns usually vary somewhat by frequency. In most cases, lower frequencies are picked up in a less-directional fashion.

3.12 IMPEDANCE OF MICROPHONES

The third factor we use to categorize microphones is **impedance**, a characteristic similar to resistance and common to audio equipment. Impedance is expressed in ohms, and mics can be either high-impedance (10,000 ohms or higher) or low-impedance (600 ohms or less). Most broadcast mics are low-impedance, since this type of mic provides the best frequency response, and most broadcast equipment is designed to accept this type of mic. High-impedance mics are also quite limited in the length of mic cable that can be used with them before hum and severe signal loss occurs. High-impedance mics should not be plugged into tape recorders or other equipment designed for low-impedance; similarly, low-impedance mics should not be used with high-impedance equipment, although the negative effects aren't as noticeable with this combination. If impedance is mismatched, sound will be distorted. There are impedance converters that can convert one type of impedance to the other.

3.13 SENSITIVITY OF MICROPHONES

Sensitivity refers to a microphone's efficiency or ability to create an output level. For the same sound source (say, one particular announcer's voice), a highly sensitive mic produces a better output signal than a less sensitive mic. To compensate for this, the gain control (volume) must be increased for the less sensitive mic; this increased gain produces more noise. Although different sensitivity-rating

systems can be employed, generally condenser mics have high-sensitivity specifications, and dynamic mics have medium sensitivity.

3.14 PROXIMITY EFFECT AND BASS ROLL-OFF

Announcer use of microphones sometimes produces a sound phenomenon known as the **proximity effect**. The proximity effect is an exaggerated bass boost that occurs as the sound source gets closer to the microphone. This should be noticeable as the announcer gets about 2 or 3 inches from the mic and is especially noticeable with mics that have a cardioid pickup pattern. Although it could help deepen a normally high voice, the proximity effect is usually defeated by a **bass roll-off switch** on the mic. This switch, when turned on, will electronically "roll off," or turn down, the bass frequencies that would be boosted by the proximity effect, thus leaving the desired flat response. Bass boost can also be controlled at the audio console, using the equalizer controls associated with the mic channel (review 2.14).

3.15 MICROPHONE FEEDBACK

Feedback is a "howling" signal generated when a sound picked up by a microphone is amplified, produced through a speaker, picked up again, amplified again, and so on endlessly. Generally, reducing the volume or turning off the mic ends the feedback. Feedback is a common mic problem in public address situations but not usually in radio production, since the speaker is muted when the mic is switched on as was mentioned in the chapter on the audio console. Occasionally, announcers can produce feedback in the production studio by operating their headphones at an excessive volume or accidentally picking up a stray speaker signal from an outside source, such as another studio speaker.

3.16 MULTIPLE MICROPHONE INTERFERENCE

Sometimes when two or more microphones receiving the same sound signal are fed into the same mixer, the combined signal becomes **out of phase**. What happens is that the sound reaches each mic at a slightly different time so that while the sound wave is up on one mic, it's down on the other. Under these circumstances, the resulting sound will have frequency peaks and cancellations causing very poor sound quality.

This situation is known as **multiple microphone interference** and can be avoided by remembering a three-to-one ratio. That is, if the mics are about one foot from the announcer (sound source), they should be at least 3 feet apart from each other. In this way the signals won't overlap (see Figure 3.14A). Another solution to this problem is to place microphones that must be close together head to head.

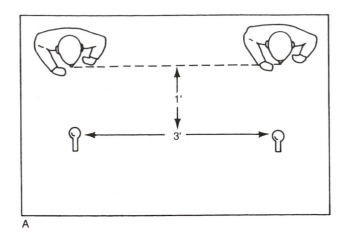

A

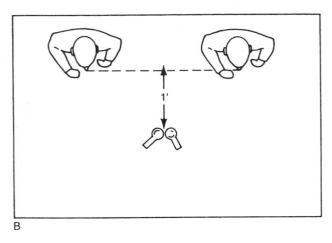

B

FIGURE 3.14 Methods of avoiding multiple-microphone interference.

Then they will get the signal at the same time (see Figure 3.14B). Although multiple microphone interference isn't usually a problem in the studio, it can occur in some radio remote situations.

3.17 STEREO

We hear sound in stereo because most sounds arrive at one ear before the other. In other words, the right ear hears a slightly different sound perspective from the left ear. This allows us to locate sound by turning our head until the sound is "centered." Most modern production work is done with stereo equipment. For example, the audio console has two master channels—a left and a right channel—and often we can pan the audio signal in such a fashion so that more of it is assigned to either the left or right channel. When we compare stereo sound to monophonic sound, we quickly realize that stereo adds both "depth" and "imaging" to the sound. *Imaging* is the apparent placement of the sound between the left and right plane and as noted provides loca-

tion of the sound in space. *Depth* is the apparent placement of the sound between the front and back plane and often provides the ambience of the space where the sound is produced. Good stereo sound allows us to record and reproduce sound as it is in real life.

3.18 STEREO MIKING TECHNIQUES

Much radio production work is done with a mono microphone. Even in "stereo" studios, usually the same mono mic signal is just sent to the left and right channels. However, there may be times when you will want to employ true stereo miking. Using traditional microphones, there are several common stereo miking techniques: **A-B** (or **spaced pair** or **split-pair**) miking, **X-Y** (or **cross-pair**) **miking**, **ORTF miking**, **stereosonic miking**, and **M-S** (or **mid-side**) **miking**. Each of these techniques is a form of **coincident miking** because they simultaneously employ multiple microphones whose pickup patterns overlap.

Since the ultimate goal of stereo recording is to have separate sound signals come from the left and right speakers, one common way to accomplish this would be to split a pair of omnidirectional mics to the left and right of center about 18 to 36 inches apart. This A-B technique uses one mic to feed the left channel of the stereo signal and another one to feed the right channel. As long as phase problems are watched and the sound source is kept a somewhat equal distance from the mics, a true stereo signal will be obtained. However, when using this technique, the sound is often "spacious," with a great deal of separation. Room ambience may be overbearing, with individual sounds seeming to "wander" in the stereo image.

Another stereo miking technique requires placing two mics (usually cardioid) like crossed swords forming an X and Y axis as shown in Figure 3.15. The angle formed between the heads of the mics is somewhere between 90 and 140 degrees, with the right microphone facing the left side of the announcer and the left mic facing the right side of the performance area. This X-Y miking allows the sound signal to reach both mics at the same time, i.e., in phase. Although this solves some of the split-pair arrangement problems, it can cause a loss of "focus" on the image because the mics are off-axis to the center of the stereo image.

ORTF technique comes from a microphone placement developed by the French broadcasting organization Office de Radio-diffusion-Television Francaise. Similar to X-Y miking, two cardioid mics are crossed so that the heads are 17 centimeters apart (6.7 inches) and form an angle of 110 degrees. Unlike X-Y miking, the right mic provides the right channel output, and the left mic provides the left channel output. ORTF miking provides a very natural stereo sound with an "openness" that surpasses typical X-Y placement.

A lesser known stereo miking technique, stereosonic miking, involves placing two bidirectional mics one on top of the other. The front of each mic faces the announcer at about a 45-degree angle, and the rear of each mic faces the back of the performance area.

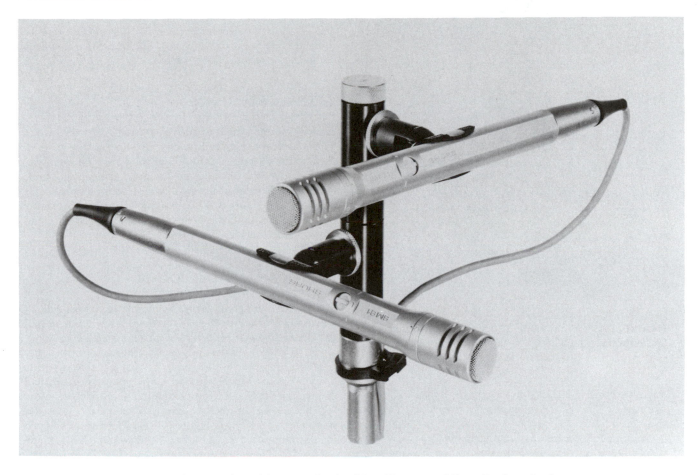

FIGURE 3.15 Cardioid microphones arranged for stereo (X-Y) miking. (Courtesy of Shure Brothers, Inc.)

Mid-side miking, or M-S miking, offers superior imaging by using mics arranged in an "upside-down T" pattern. The mid mic (often a supercardioid) is aimed at the sound source and the side mic (usually a bidirectional) is placed at 90 degrees to the mid mic to pick up the sound to the left and right. Both microphones must be fed into a mixer designed to *matrix*, or decode, the incoming signals into a stereo signal.

Variations of these techniques, using different types of microphones and mic placements, are also employed to achieve stereo miking. Also, stereo microphones can duplicate X-Y or M-S techniques using a single microphone with specially designed internal sound-generating structures.

In addition to the basic microphone, the radio production person will find one or more of the following accessories necessary for proper audio production work: windscreens, shock mounts, and stands or booms.

3.19 WINDSCREENS

The most common **windscreens** (sometimes referred to as **blast filters** or **pop filters**) are ball-shaped accessories that can be placed over the head or front of the microphone to help reduce the chance of a plosive sound. They can also be built into the grill of the microphone as shown in Figure 3.16. Another design consists of a porous, filmlike material that is suspended within a circular frame. This windscreen

Built-in Pop Filter

FIGURE 3.16 Microphone with pop filter. (Courtesy of Shure Brothers, Inc.)

FIGURE 3.17 Microphone shock mount. (Courtesy of Shure Brothers, Inc.)

apparatus is then attached to the microphone stand and positioned in front of the microphone rather than placed on the mic.

Announcing words that emphasize the *p*, *b*, or *t* sound naturally produces a sharp puff of air and can produce a pop or thump when hitting the mic (especially when a dynamic mic is closely worked). The pop filter, however, eliminates or reduces this problem.

3.20 SHOCK MOUNTS

A **shock mount** is often employed to isolate the microphone from any mechanical vibrations, or shocks, that may be transmitted to the mic through its stand. The mic is physically suspended (usually by an arrangement of elastic bands) and isolated from the stand or boom to which it is attached (see Figure 3.17). If the mic stand is accidentally bumped, the sound of this thud will not be passed on and amplified by the mic.

3.21 MIC STANDS AND BOOMS

Mic stands consist of two chrome-plated pipes, one of which fits inside the other. A rotating clutch at one end of the larger-diameter pipe allows the smaller pipe to be adjusted to any height desired. At the other end of the larger pipe is a heavy metal base (usually circular) that supports the pipes in a vertical position. Some mic stands utilize a single pipe that's at a fixed height. In either case, the mic is attached to the top of the pipe by a standard thread and mic-stand adapter (see Figure 3.18A). **Floor stands** adjust for the announcer in a standing position, and **desk stands** are used for the seated announcer.

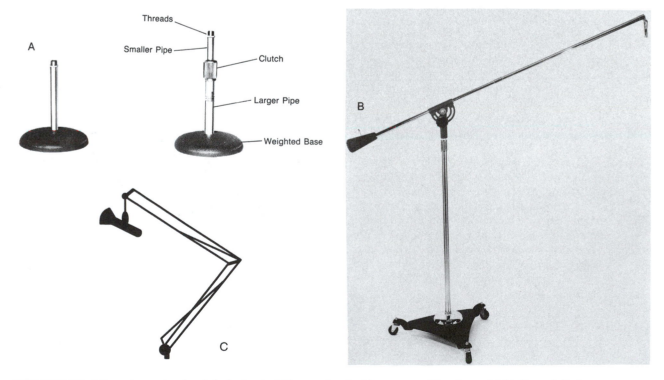

FIGURE 3.18 Microphone stands: a) desk stands; b) boom stand; c) boom arm. (Courtesy of Atlas/Soundolier)

A **boom stand** is a long horizontal pipe that attaches to a floor stand (see Figure 3.18B). One end of the boom is fitted with the standard thread (for the mic), and the other end is weighted to balance the mic. The horizontal pipe allows the boom stand to be away from the announcer and positions the mic in a workable relationship to him or her. A **boom arm** (see Figure 3.18C) is a mic stand especially designed for use in the radio studio. It consists of metal rods and springs designed somewhat like a human arm; the mic attaches to one end, and the other end goes into a mounting base. The whole unit can then be attached to production studio furniture so that the mic can be placed in close proximity to the audio console.

3.22 MICROPHONE USAGE

An understanding of the microphone types, pickup patterns, and other mic characteristics is only useful if you can apply this knowledge to everyday radio production use. It seems reasonable to assume that the microphone in a broadcast studio will be a dynamic mic because that's the most commonly used mic in broadcasting, but it could also be a condenser mic. Perhaps the more important consideration is its pickup pattern. In the studio, we mainly want to pick up the announcer's voice and some ambience of the studio. A cardioid pickup pattern, with its pickup of front and side sounds, works best to accomplish this and not pick up unwanted sounds from the rear, such as the announcer moving papers or turning on and off various switches.

When using a microphone in the standard studio setup, keep in mind two basic rules concerning distance from the microphone and position of the microphone. The announcer's mouth should be about 6 inches away from the mic. That's about the length of a dollar bill—a good way to remember mouth-to-mic distance. This is a good starting point for using a mic. You may find that you need to be closer or farther away because of the strength of your voice or the vocal effect that you're trying to achieve. Position the microphone so that you are not talking directly into it. Put the mic level with your nose and tilted down toward your mouth, and talk beneath the front of the mic or place the mic below your mouth with the front of the microphone tilted up toward your mouth, allowing you to talk across the top of the mic.

In radio production work, it's not uncommon to leave the studio to record an interview. Certainly the radio news person does this frequently. In a remote location interview, the dynamic microphone would probably be used because of its ruggedness and good quality. In most cases, at a remote site you *want* to pick up the ambience of the location and various voices (particularly the announcer and the interviewee). The omnidirectional pickup pattern, with its ability to pick up sound equally well in all directions, provides a solution for this situation. If there's likely to be a great deal of background noise, a cardioid mic might be better, but it will have to be carefully placed so that it picks up both voices well.

Interviews (and some other broadcast situations) may require you to mic a group of people in the studio seated around a table. You could use a single omnidirectional mic in the center of the table, or you could use a separate cardioid mic for each individual at the table. Another option is to use the PZM, or pressure zone microphone.

Sportscasters are commonly out of the studio to broadcast games and other sports events. The sportscasters' headset provides both earphones and a microphone in a single unit, leaving the announcer's hands free (see Figure 3.19). The mic is usually a dynamic mic with either a cardioid or an omnidirectional pickup pattern. If a cardioid mic is employed, the sportscaster may want to have an additional mic for picking up crowd and sports action noise.

3.23 CONCLUSIONS

We said earlier that there is no one correct microphone to use in radio production. When using mics in any production, don't be afraid to experiment in each particular situation. The bottom line in any production should be how it sounds. It's good to be flexible in radio production because often you won't have a wide variety of microphones available. If you use what you've learned in this chapter, you should be able to obtain clear appropriate sound under various circumstances.

Headphone Section

Microphone Section

FIGURE 3.19 Sportscasters' headset. (Courtesy of Telex Communications, Inc.)

Self-Study

■ QUESTIONS

1. Another name for the dynamic microphone is _____.
   a) condenser
   b) pressure
   c) capacitor
   d) PZM

2. The dynamic mic's sound-generating element is constructed of a diaphragm, permanent magnet, and some coils of wire wrapped around the magnet. Into which of these is a small electrical current induced during use?
   a) diaphragm
   b) magnet
   c) coil
   d) none of the above

3. The condenser mic differs from the dynamic mic in that _____.
   a) the condenser mic needs a power supply, and the dynamic mic doesn't
   b) the dynamic mic has a diaphragm, and the condenser mic doesn't
   c) the dynamic mic has better sound quality than the condenser mic
   d) the condenser mic is bidirectional, and the dynamic mic is omnidirectional

4. Which kind of mic is most likely to be found in a broadcast-quality audio cassette recorder?
   a) dynamic
   b) ribbon
   c) wireless
   d) condenser

5. The mic that picks up sound on all sides is _____.
   a) hypercardioid
   b) cardioid
   c) omnidirectional
   d) bidirectional

6. Which mic would be most appropriate for conducting an interview on the sidelines at a football game?
   a) unidirectional
   b) omnidirectional
   c) bidirectional
   d) hypercardioid

7. Which mic would be most appropriate for picking up a sportscaster announcing at a baseball game?
   a) cardioid
   b) nondirectional
   c) bidirectional
   d) omnidirectional

8. Most broadcast-quality microphones are _____.
   a) low cost
   b) high-impedance
   c) low-sensitivity
   d) low-impedance

9. Which of the following is most likely to exaggerate the bass sounds of a person's voice?
   a) feedback
   b) proximity effect

     c) multiple-microphone interference
     d) bass roll-off

10. The purpose of a shock mount is _____.
     a) to reduce plosive sounds
     b) to keep the announcer's head at least 12 inches away from the mic
     c) to isolate the microphone from mechanical vibrations
     d) to prevent static electricity discharges

11. Which type of mic stand can be farthest away from a person and still allow the person to be close to the microphone?
     a) boom stand
     b) floor stand
     c) desk stand
     d) shock mount

12. Which stereo miking technique uses two microphones crossed (like swords) at a 90-degree angle to each other?
     a) split-pair miking
     b) mid-side miking
     c) M-S miking
     d) X-Y miking

13. Which type of microphone uses small multiple sound generating elements within a single mic housing?
     a) ribbon mic
     b) wireless mic
     c) stereo mic
     d) boundary mic

14. A microphone's pickup pattern is exactly the same thing as a microphone's polar response pattern.
     a) true
     b) false

15. Which microphone has a permanently charged capacitor as part of its sound-generating element?
     a) condenser mic
     b) regulated phase mic
     c) electret mic
     d) ribbon mic

16. Which microphone would be best at picking up sound when the sound source was a considerable distance from the mic?
     a) lavalier mic
     b) PZM mic
     c) RF mic
     d) shotgun mic

17. Sound picked up from the rear of a microphone is 90 degrees off-axis.
     a) true
     b) false

18. Which microphone pickup pattern describes a mic with the narrowest scope of on-axis sound pickup?
     a) cardioid
     b) ultracardioid
     c) hypercardioid
     d) supercardioid

19. Microphone impedance refers to a microphone's ability to create an output signal.
     a) true
     b) false

20. A "screech" that occurs when sound is picked up by a microphone, amplified, fed back through a speaker, and picked up again over and over is called _____.
   a) proximity effect
   b) multiple-microphone interference
   c) feedback
   d) bass roll-off

21. Which of the following is *not* a way that microphones are commonly classified?
   a) a mic's sound-generating element
   b) a mic's size
   c) a mic's pickup pattern
   d) a mic's impedance

22. Which stereo miking technique requires exact placement of two cardioid microphones?
   a) ORTF miking
   b) Stereosonic miking
   c) X-Y miking
   d) A-B miking

23. Which microphone has a wire spiral imbedded in a circular diaphragm as part of its sound-generating element?
   a) dynamic mic
   b) electret mic
   c) moving coil mic
   d) regulated phase mic

24. Multiple microphone interference can be avoided if the mics are at least three times as far from each other as they are from the sound source.
   a) true
   b) false

25. As a final test on microphones, match the items in the top list (1, 2, 3 . . .) with the choices in the bottom list (o, a, c . . .), and then select the correct set of answers from the sequences shown in a, b, c, or d below.

   1. _____ cardioid
   2. _____ dynamic
   3. _____ high impedance
   4. _____ feedback
   5. _____ sensitivity
   6. _____ capacitor
   7. _____ omnidirectional
   8. _____ pop filter
   9. _____ wireless mic
   10. _____ polar pattern
   11. _____ proximity effect
   12. _____ split-pair
   13. _____ ORTF miking
   14. _____ PZM

   o. a mic that picks up from all directions
   a. a mic that picks up from all but one direction
   c. another name for a condenser mic
   d. a microphone with a diaphragm, magnet, and coil
   w. another name for a windscreen
   p. a mic, often mounted on a surface, that is also known as a plate mic
   s. a howling sound
   h. a mic that cannot have a very long cable without developing hum
   e. a microphone's efficiency
   b. can be eliminated by a bass roll-off switch

m. a form of stereo miking that uses two omnidirectional mics
k. a form of stereo miking that uses two cardioid mics
t. a two-dimensional drawing of a mic response pattern
r. a mic that uses a low-power transmitter and receiver

a) 1.a 2.d 3.h 4.s 5.e 6.c 7.o 8.w 9.r 10.t 11.b 12.m 13.k 14.p
b) 1.o 2.p 3.h 4.s 5.e 6.d 7.a 8.w 9.r 10.t 11.b 12.m 13.k 14.c
c) 1.a 2.d 3.s 4.h 5.w 6.c 7.o 8.e 9.p 10.b 11.r 12.k 13.m 14.t
d) 1.t 2.c 3.b 4.h 5.r 6.d 7.o 8.s 9.k 10.a 11.p 12.e 13.m 14.w

■ ANSWERS

If You Answered A:

1a. No. This is a different type of microphone. (Reread 3.3 and 3.4.)
2a. Wrong. The diaphragm feels the pressure. (Reread 3.3.)
3a. Right. The condenser mic power supply is needed to charge the backplate and diaphragm.
4a. No. While the dynamic mic is often found in the broadcast studio, it is not the correct response here. (Reread 3.3 and 3.4.)
5a. No. Hypercardioid mics pick up mainly from the front. (Reread 3.7–3.10.)
6a. Wrong. As a cardioid mic, it picks up in mainly one direction, so it would not pick up much of the crowd ambience. (Reread 3.7–3.10.)
7a. Yes. It would pick up the sportscaster without much of the background noise.
8a. No. Most broadcast studio mics cost $300–$500. While remote mics may be less, they are usually not inexpensive. (Reread 3.12.)
9a. No. Feedback is a howling noise caused by having open mics near speakers. (Reread 3.14 and 3.15.)
10a. No. A pop filter is used to reduce plosive sounds. (Reread 3.19 and 3.20.)
11a. Right. This is the best selection.
12a. No. This is a different stereo miking technique. (Reread 3.18.)
13a. No. As the name implies, a ribbon mic employs a thin metallic ribbon as part of a single sound-generating element. (Reread 3.5 and 3.6.)
14a. Wrong. A pickup pattern is the three-dimensional shape around the mic in which it hears the sound best; a polar response pattern is the two-dimensional representation of this. (Reread 3.11.)
15a. You're close to the correct response, but normal condenser mics require a power supply to charge the capacitor. (Reread 3.4.)
16a. Wrong. Lavalier mics are designed to be attached to the announcer's clothing. (Reread 3.6.)
17a. No. Sound that is 90 degrees off-axis would be coming from the side of a microphone. (Reread 3.7.)
18a. All cardioid microphones pick up sound mainly from the front, but this is not the correct answer. (Reread 3.10.)
19a. This is not correct. Impedance is an electrical characteristic similar to resistance. (Reread 3.12 and 3.13.)
20a. No. Proximity effect is a characteristic of microphones that accents the bass response of the mic. (Reread 3.14 and 3.15.)
21a. No. Microphones are categorized by their sound-generating elements. (Reread 3.2.)
22a. Correct. The mic heads are spaced 6.7 inches apart and form a 110-degree angle.
23a. Wrong. Dynamic mics do have a wire coil as part of their sound-generating element, but it's not configured like this. (Reread 3.3 and 3.5.)
24a. Yes. This is a true statement.
25a. Correct. You have now finished the section on microphones.

If You Answered B:

1b. Correct. The dynamic mic is also called the pressure mic or the moving-coil mic.
2b. Wrong. The magnet sets up the field. (Reread 3.3.)
3b. No. Both mics have a diaphragm. (Reread 3.3 and 3.4.)
4b. No. This is a professional-quality mic that is not used much anymore. (Reread 3.3, 3.4, and 3.5.)
5b. Wrong. The cardioid picks up sound on all but one side—usually the one right behind the mic. (Reread 3.7–3.10.)
6b. Correct. It would pick up from all sides, thus easily miking the interview and some crowd noise.

7b. No. The crowd noise would tend to drown out the announcer. (Reread 3.7–3.10.)

8b. No. Consumer-quality mics are usually high-impedance, but not broadcast quality. (Reread 3.12.)

9b. Right. When an announcer gets too close to the mic, the bass may be exaggerated.

10b. No. For one thing, the announcer's head should be about 6 inches away, not 12 inches away. (Reread 3.20.)

11b. No. A person must stand right beside a floor stand. (Reread 3.21.)

12b. No. This is a different stereo miking technique. (Reread 3.18.)

13b. No. Wireless mics employ a transmitter (part of the mic) and a receiver, but this is not a description of this system. (Reread 3.6.)

14b. Right. This is the correct response.

15b. No. The regulated phase microphone doesn't require phantom power, nor does it have a capacitor as part of its sound-generating element. (Reread 3.4 and 3.5.)

16b. Wrong. PZM mics are designed to be placed on a flat surface, such as a tabletop. (Reread 3.6.)

17b. Yes. This statement is false. Sound that is picked up from the rear of a microphone is 180 degrees off-axis.

18b. This is the correct response.

19b. This is the right answer because mic sensitivity refers to a microphone's ability to create an output level.

20b. No. This is not correct. Multiple microphone interference is a phase problem that creates peaks and cancellations in the sound. (Reread 3.15 and 3.16.)

21b. Correct. While microphones come in a wide variety of sizes, they are not usually categorized by size.

22b. Wrong. For one thing, stereosonic miking utilizes bidirectional mics, not cardioid mics. (Reread 3.18.)

23b. No. The electret mic is a type of condenser mic that uses a capacitor as its sound-generating element. (Reread 3.4 and 3.5.)

24b. You're incorrect. The statement is true. (Reread 3.16.)

25b. No. You are confused regarding the sound-generating elements and the pickup patterns. (Reread 3.1–3.11.)

If You Answered C:

1c. No. This is another name for the condenser mic. (Reread 3.3 and 3.4.)

2c. Correct. The current is in the coil.

3c. No. The condenser mic has slightly better quality than the dynamic mic. (Reread 3.3 and 3.4.)

4c. No. You're way off base here. (Reread 3.4 and 3.6.)

5c. Yes. An omnidirectional mic picks up on all sides.

6c. Wrong. It could pick up the interview, but wouldn't get much crowd noise. (Reread 3.7–3.10.)

7c. No. A bidirectional mic picks up on two sides, and the sportscaster would only be on one side. (Reread 3.7–3.10.)

8c. No. Condenser mics are high sensitivity, and dynamic mics are medium sensitivity. Neither are low sensitivity. (Reread 3.12 and 3.13.)

9c. No. This will create a distorted signal. (Reread 3.14 and 3.16.)

10c. Yes. This is a special type of microphone holder that suspends the mic.

11c. No. A desk stand has to be right in front of the person. (Reread 3.21.)

12c. No. This is a different stereo miking technique. (Reread 3.18.)

13c. Yes. This is the correct answer.

15c. Yes. A type of condenser mic, the electret mic, has a capacitor that is charged during manufacture.

16c. Wrong. RF mics are wireless mics, but they don't have exceptional distance pickup characteristics. (Reread 3.6.)

18c. While the hypercardioid does have a narrow, unidirectional pickup pattern, this is not the best response. (Reread 3.10.)

20c. Correct. That screeching, howling sound is feedback.

21c. No. Microphone pickup patterns are used to categorize mics. (Reread 3.2.)

22c. Wrong. This is a similar miking technique, but there are distinct differences. For example, the angle formed by crossed mics can vary from 90 to 140 degrees with X-Y miking. (Reread 3.18.)

23c. No. This is not correct. The moving coil mic is just another name for the dynamic mic. (Check the response to 23a; reread 3.3 and 3.5.)

25c. Wrong. You are confused about many of the elements of microphones. (Reread from 3.12 to the end of the chapter.)

If You Answered D:

1d. No. A PZM mic does not indicate a type of construction. (Reread 3.3 and 3.6.)

2d. Wrong. (Reread 3.3.)

3d. No. Both dynamic and condenser mics can have omnidirectional or bidirectional pickup patterns. (Reread 3.3, 3.4, 3.8–3.10.)

4d. Yes. Most broadcast quality cassette recorders have a built-in condenser mic.

5d. No. A bidirectional mic picks up sound from the front and back of the mic. (Reread 3.7–3.10.)

6d. No. This is a type of cardioid mic with a very narrow one-direction pickup pattern. While it might be useful to pick up one specific section of the crowd, it wouldn't provide full background crowd noise. (Reread 3.7–3.10.)

7d. No. This is another term for nondirectional. (Reread 3.7–3.10.)

8d. Right. Low-impedance is what most broadcast equipment is.

9d. Wrong. Bass roll-off is an electronic "turn down" of bass frequencies. (Reread 3.14.)

10d. No. This is not the correct answer. (Reread 3.20.)

11d. No. While this is often used in conjunction with a mic stand, it won't determine the distance between announcer and mic stand. (Reread 3.21.)

12d. Correct. This is a description of X-Y stereo miking technique.

13d. No. Boundary mics employ a single sound-generating element in conjunction with a sound-reflecting plate. (Reread 3.6.)

15d. No. The ribbon mic doesn't require a power supply, nor does it have a capacitor as part of its sound-generating element. (Reread 3.4 and 3.5.)

16d. Right. Shotgun mics are designed with a narrow tube that can be aimed at and will pick up sound from some distance away.

18d. While the supercardioid does have a narrow, unidirectional pickup pattern, this is not the correct response. (Reread 3.10.)

20d. No. Bass roll-off is a switch on some microphones that electronically turns down the lower frequencies. (Reread 3.14 and 3.15.)

21d. No. Mic impedance is used to categorize microphones. (Reread 3.2 and 3.12.)

22d. Not correct. For one thing, A-B miking uses omnidirectional microphones. (Reread 3.18.)

23d. This is the correct response.

25d. Wrong. You have confused a number of important terms. (Reread the entire chapter.)

Projects

■ PROJECT 1

Position microphones in various ways to create different effects.

Purpose

To enable you to experience proximity effect, feedback, multiple-mic interference, and the differences in sound quality that occur when a microphone is placed at different distances and angles from an announcer.

Advice, Cautions, and Background

1. You have not read the chapter on audio tape recorders yet, so you may need some help from your instructor or the engineer in setting up the equipment.

2. For the proximity effect, try to find a mic that doesn't have a bass roll-off switch, or make sure it's switched off if it does.

3. Don't allow feedback to occur for too long. It can be damaging to all the electronic equipment—and your ears. You may have to plug the mic into something other than the audio board if the board automatically shuts off the speakers when the mic is turned on.

4. For the multiple-mic interference, make sure the mics are closer than 3 feet. Omnidirectional mics will demonstrate the effect the best.

5. Use a cardioid mic for the distance and angle experiments because it will demonstrate the points better.

How to Do the Project

1. Set up an audio board so that two microphones are going into it, and the sound of those two mics can be recorded on a tape recorder.
2. Put the tape recorder in record and activate one of the microphones. Start talking about 2 feet away from the mic, and keep talking as you move closer until you are about 2 inches from it. As you talk, say how close you are to the mic, and mention that you're experimenting with the proximity effect. Stop the tape recorder.
3. Position a microphone so that it's close to an activated speaker. Turn on the tape recorder, and talk into the mic. Record a short amount of the feedback, and turn off the tape recorder.
4. Position two mics in front of you that are less than 3 feet apart. Put the tape recorder in record, and talk into the microphones, saying that you're testing for multiple-microphone interference. Turn off the tape recorder.
5. Position one mic in front of you. Put the tape recorder in record. Position yourself 12 inches from the mic and talk into it. Then position yourself 6 inches from the mic and talk directly into it. Then talk across the top of the mic. Move 6 inches to the side of the mic, and talk into it with your mouth positioned to speak across the top of it. Get behind the mic, either by moving behind it or by turning it around, and talk from about 6 inches away. Describe each action as you do it.
6. Listen to the tape to hear the various effects and to see that you have, indeed, recorded all the assignments. If some of them didn't turn out as well as you would have liked, redo them.
7. Turn in the tape to your instructor for credit for the project. Make sure you put your name on the tape and label it MICROPHONE PLACEMENT.

■ PROJECT 2

With several other students, make a recording using stereo miking techniques.

Purpose

To give you experience utilizing standard microphones to employ various stereo miking techniques.

Advice, Cautions, and Background

1. Which technique you employ will depend on the microphones that are available at your facility.
2. Since this project requires several people, your instructor may assign it as a class project.
3. The "How To" section that follows uses X-Y techniques for illustration purposes; you can adjust it for any of the other techniques.

How to Do the Project

1. At one end of your studio (or room) arrange at least three students in a "left"–"center"–"right" configuration. This could also be three groups of students or even a small musical group, as long as they're arranged so that specific sections can be identified as being left, center, or right.
2. Set up two cardioid microphones to record onto an audio tape recorder.
3. Arrange the microphones in an X-Y position as described in the text, in line with the "center" of the students or group.
4. Begin recording and have the students or group do the following:
   a. have the "left" say something
   b. have the "right" say something
   c. have the "center" say something
   d. have all three sections say something different, but at the same time
   e. have just the "center" say something
   f. have the "left" and "right" say something
   g. have all three sections say something
5. As you are recording the above, make sure to identify what each segment is by having one student say something like, "This is just the 'left' group," before the group says anything.

6. Rewind and play back the tape. Are you able to "hear" the stereo image correctly? Does the left group "appear" to be located left? Make some observations about the recording results and write them down.
7. Repeat the same recording, but use a single microphone; i.e., record in mono.
8. Listen to the second recording. What differences do you hear between the two recordings? Write down your observations again.
9. Turn in your tape and the observation sheet labeled STEREO MIKING to your instructor to receive credit for this project.

■ PROJECT 3

Compare sound from different types of microphones.

Purpose

To enable you to know the differences between dynamic and condenser mics and between cardioid and omnidirectional mics.

Advice, Cautions, and Background

1. You have not read the chapter on audio tape recorders yet, so you may need help from your instructor or engineer in setting up the equipment.
2. Don't be concerned if you can't discern differences between the dynamic and condenser mics. The pop and hiss effect doesn't occur for all voices.
3. If your facility has mics with other sound-generating elements (such as ribbon or regulated phase) or other pick-up patterns (such as bidirectional or ultracardioid), add those to the exercise.

How to Do the Project

1. Select a dynamic mic and a condenser mic from those at your facility.
2. Attach the mics to a tape recorder. You can put both mics through an audio console and use them one at a time, or you can attach each mic into the tape recorder in its turn.
3. While you're recording, say the following into the dynamic mic: "Peter Piper picked a peck of pickled peppers," and "She sells seashells by the seashore." Turn the recorder off.
4. Repeat step 3, but use the condenser microphone.
5. Select a cardioid mic and an omnidirectional mic from those at your facility.
6. Attach the microphones into a tape recorder as in step 2.
7. Position the cardioid mic on a stand, and, while recording, walk totally around the mic, saying where you are as you move; e.g., "Now I'm right in front of the mic; now I'm ninety degrees to the right of the front of the mic." Turn the recorder off.
8. Repeat step 7, using the omnidirectional microphone.
9. Listen to the tape, and write down any observations you have about the differences among the various mics.
10. Turn in the paper and tape to your instructor for credit for the project. Be sure to include your name on both and call the assignment MICROPHONE COMPARISONS.

CHAPTER FOUR

CD Players (and Turntables)

Information

4.1 INTRODUCTION

In an audio production facility, the CD player is one of the main sources for playing back prerecorded material. Because CDs are often aired one right after the other, there are usually at least two CD players in each production room or on-air studio. In this way, one can be cued while the other is playing on the air.

Since the advent of the compact disc, turntables have steadily lost favor as a necessary piece of production equipment. While a few facilities have no turntables at all, most still employ the turntable to a limited extent. For example, some small stations still play records on the air, and some stations use turntables for both on-air and production room playback of specialty material that is not available on CD or tape. Of course, one of the main reasons the CD player displaced the turntable is the CD player's ability to provide digital sound.

4.2 DIGITAL VERSUS ANALOG

In the radio production domain, digital technology is replacing analog technology. By digital technology we mean the process of converting audio information into an electrical signal composed of a series of on and off pulses. In other words, the digital process is a conversion into **binary** numbers. Computers handle information in this manner by associating a binary number with each letter of the alphabet and each number, then manipulating this binary data.

Before digital technology was developed, audio recording relied on an **analog** process. An analog signal (for example, the audio signal produced by a microphone) is a continuously variable electrical signal whose shape is defined by the shape of the sound wave produced (see Figure 4.1). In the analog recording process, a duplicate, or electromagnetic representation, of the sound wave of the original sound source is stored on magnetic tape.

Each time the analog signal is recorded or processed in some fashion, it is subject to degradation because the signal changes shape slightly. Analog encoding is similar to creating a line graph to show a statistical analysis. All measurements are on a continuous line that curves up and down with no discrete points. The recording process is like trying to retrace a curve on a graph; the reproduction is always slightly different than the original.

In addition to this generation loss, because analog recording relies on magnetic pulses stored on tape, any defect or decrease of the magnetic properties of the tape means a loss of signal quality. Typical problems for analog recording have included noise and distortion, print-through and crosstalk, flutter and hiss, and limited dynamic range. You'll learn more about these problems in the next chapter on audio tape recorders.

On the other hand, digital encoding is accomplished in a discrete fashion, like looking at individual numbers in a statistical analysis and writing them down in a set order. The audio signal (from a microphone) starts out as analog, but it can be converted to digital by going through four stages: filtering, sampling, quantizing, and coding. First, the original sound signal is sent through a low-pass **filter** that "strips" off frequencies that are above the range of human hearing. Although originally inaudible, these frequencies can be shifted, or aliased, into an audible range during recording and playback, so this process is known as anti-aliasing.

The filtered analog signal is next divided many times a second in a process known as **sampling** (see Figure 4.2). By taking enough samples and converting the samples to binary data, an exact reproduction of the original sound signal can be recorded onto tape. Digital equipment utilizes **sampling rates** of 32, 44.1 and 48 thousand times per second. However, many pieces of digital equipment are designed to oversample at a rate up to eight times the basic rate for even better sound quality and accurate reproduction.

Quantizing and coding are the stages that assign a numerical value to each individual sample. **Quantizing** determines how many levels or values the sample will be broken down into. Audio engineers use the term "bit" to reference quantizing levels; the more levels, the more information you would have about the signal. For example, a 1-bit system would signify just two quantizing levels—either there's a voltage or there isn't—not much information about the sound. Each additional bit doubles the number of levels. The standard bit rate for most digital recording is 16 bits, or 256 quantizing levels, and some compact disc recording is now being done with a 20-bit technology.

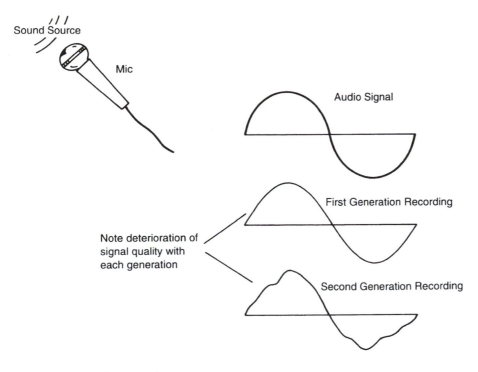

FIGURE 4.1 Analog recording.

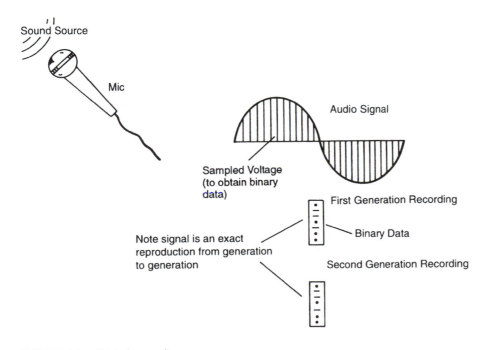

FIGURE 4.2 Digital recording.

Coding is putting sixteen "0's" and "1's" in a precise order corresponding to the values measured during the quantizing process. This binary, or digital, "word" represents each individual sample and is a voltage level that indicates the loudness and timbre information of the sound at that particular moment.

Remember, it's the binary data that is actually recorded on the tape, not an analog representation of the signal. With digital technology, we can copy from tape to tape with no measurable loss of quality. Along with the improved frequency response, wide dynamic range, and unmeasurable noise and distortion, this ability to rerecord with no decrease in quality has been a big part of digital's acceptance in radio production.

While most of the problems associated with analog recording are nonexistent with the digital process, digital recording does have a unique, subjective problem of its own. The superior sonic quality of a digital signal is so clear and noise free that some find it sounds "harsh" or "mechanical" in comparison to the old "warmth" of analog distortion!

4.3 COMPACT DISCS

Until some other technology develops, the **compact disc** (**CD**) has essentially replaced the vinyl record, and the CD player has replaced the turntable in the radio studio. The CD is a small plastic disc 4.7 inches in diameter. On one side of the disc, the music is stored as a spiral of microscopic pits and lands (or flat areas) that contain the encoded information about the sound. In addition to music information, data encoded on the disc tells the CD player where each track begins and ends, how many tracks are on the disc, and other timing and indexing information. A CD stores music as 16-bit digital words (or 16 consecutive zeros or ones), going from the center hole to the outside rim of the disc. With the digital sampling rate, you get 44,100 of these digital "words" for each channel every second! In fact, there are over 5 billion bits of data stored on a CD just to produce about 60 minutes of music. A thin aluminum reflective coating (some CDs use a gold-coated reflective surface) less than 70 millionths of a millimeter makes it possible for the **laser** light to read the encoded data, and a layer of clear polycarbonate plastic protects the encoded information. The reverse side of the CD is made up of the label and a protective lacquer coating (see Figure 4.3).

CDs were originally designed to hold 60 minutes of recorded material; however, some have been released with close to 75 minutes of music. CD singles (3-inch CDs) have also been developed and are capable of holding about 20 minutes of music. The number of tracks and timing information about songs have always appeared on the display of CD players, but why not other information, such as song titles? While it's been technically possible to do so, such provisions were not put into the CD "standards." Co-developers of the compact disc, Sony and Philips, have agreed to revise the CD standard so that information about the artists, albums, and songs can be included with the music. Of

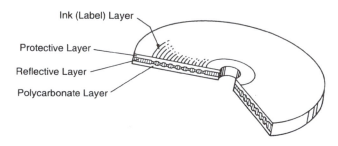

FIGURE 4.3 Compact disc cutaway.

course, a text-capable audio CD player will be required, but they may be on the market in the near future. Computer CD-ROMs *are* text-capable players, and some music CDs now contain additional information that can be accessed with a PC.

The Society of Professional Audio Recording Services (SPARS) developed a code to put on CD packaging to show just how "digital" the CD was. The code told whether the recording was from a digital or analog master and whether it was mixed and mastered in analog or digital. Many AAD or ADD coded CDs didn't appeal to digital purists looking for that completely digital CD, labeled DDD, so the code was dropped for a few years. However, in 1995, SPARS recommended the code's return, and many of today's CDs again show the code.

The next generation of compact disc may well be the **Digital Versatile Disc** (**DVD**) or a variation of it. Originally designated as Digital Video Disc, the DVD was mainly designed to hold a feature-length movie on a compact disc-sized medium. Primarily a new video format, it can also be used for computer data or music and will probably have applications in both areas. Physically, the DVD is the same size as a compact disc (about 5 inches); however, it has two data layers that can both be read from one side of the disc by refocusing the laser in the DVD player. The data information is spaced much closer than on a compact disc, so by using a data compression scheme and a two-sided disc, the DVD could hold up to 17 gigabytes of data, or the content of about 25 current compact discs! It's being suggested that the audio specifications must be even better than CD quality. In addition to massive storage capacity, an audio-only DVD could have a sampling rate of 96 mHz, 24-bit quantizing, and a dynamic range of more than 120dB. As of this writing, the final standards for the DVD are being worked out, and the implications of DVD in the audio area are uncertain.

4.4 CARE OF CDS

The compact disc was originally promoted as indestructible, but practical use in the radio studio has shown that they do require some care. Some CD players can be mistracked by dust or fingerprints on the CD surface, and serious scratches can also render the CD unplayable.

CDs can be cleaned by wiping the aluminum side (opposite the label) with a lint-free cloth gently from the center hole of the CD directly out toward the CD rim. Do *not* use a circular cleaning motion because this could cause a scratch that follows the spiral data on the disc, causing the disc to mistrack.

When removing a CD from its jewel box case, put your thumb on the center spindle of the jewel box and your index or middle finger on the rim of the disc. Press down with your thumb and gently pull up on the disc until it is free of the case. When handling a disc, continue to hold only the outer edges or inside hole, and avoid bending the CD in any manner. Oily fingers or flexing a disc can damage the CD's protective coating and ultimately allow oxidation of the aluminum layer, which can ruin the CD.

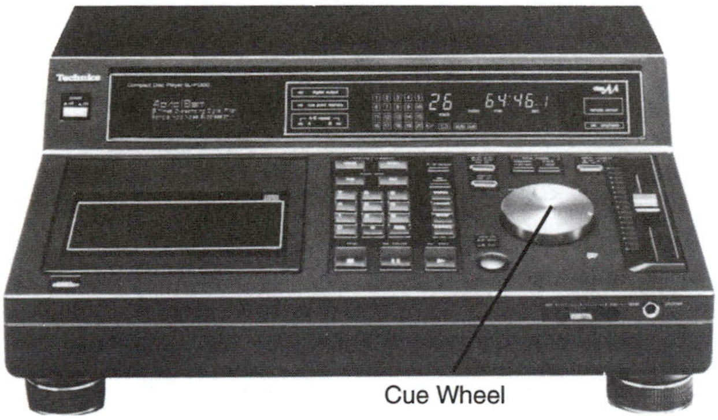

Cue Wheel

FIGURE 4.4   Professional-quality CD player. (Courtesy of Panasonic Company)

4.5   THE CD PLAYER

The CD player shown in Figure 4.4 is typical of a unit designed specifically for broadcast use. Many radio stations and production facilities just use consumer CD players, but these often require special interfaces to be compatible with other broadcast equipment and are not designed for daily use. Good production practice dictates the use of broadcast-quality CD players rather than consumer models.

One of the main features of a broadcast-quality unit, and its main difference from a consumer unit, is a large **cue wheel** that allows the operator to "rock" the CD back and forth, much like cueing a record. In this manner the exact start point of the music can be found, and then the player can be paused at that spot, ready for play. Similar systems

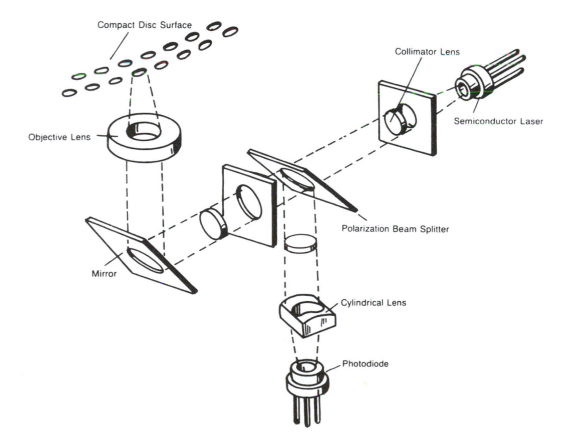

FIGURE 4.5   Simplified internal structure of a CD laser.

use a single microprocessor control unit that allows cueing of two separate playback units. Other normal CD controls include a play button to start the CD, several controls to select specific *tracks* (or songs) on the CD, and an open/close control to load the CD into the machine. Regardless of the exact design, the CD ends up in a **tray** or **well** (see Figure 4.7), where it spins so the laser can read it.

The internal structure of the CD player centers around the laser beam system that reads the encoded data on the surface of the CD. Figure 4.5 shows a simplified drawing of this optical pickup. A **laser diode** generates the laser light beam, and a **prism system** directs it toward the disc surface. Different types of lenses focus the laser beam exactly on the pits of data. Reflected light is directed back through the lens and prism to a wedge lens and a **photodiode**, which provides the data signal that will be converted to an audio signal.

Some CD players utilize a single-beam laser, and some employ a three-beam pickup. There is no sound quality difference between them. Actually, all CDs employ a single laser, but the three-beam players track the CD by splitting the laser light into three side-by-side spots. The middle spot picks up the data, and the outer spots keep the laser on track by focusing on the blank space between the data tracks. If an outer spot begins to pick up audio data, the player knows it is off track and can correct it. In a single-beam design, if the laser wanders off track, a difference in the intensity on one side of the reflected light is detected, and the player can correct it. The three-beam pickup is more complex, and some manufacturers favor the single beam approach.

A different design approach to CD players has been to build them so that the CD must be put into a plastic housing before it can be played in the unit (see Figure 4.6). As you'll see in the next chapter, this is more similar to the audio tape cartridge than to the turntable design previously used. Special broadcast features of this type of CD player include the ability to select one track while another is playing (so two songs can be played back-to-back from the same CD) and a countdown timer (so the operator knows exactly how much time remains on the song currently playing). The plastic cartridge increases the cost of this system, but it also affords the CD extra protection from dirt and damage.

As the radio station CD library continues to grow, some use has developed for a professional **multiplay** CD player. With an ability to hold up to 200 CDs and with interfacing capability, this unit has been used as a music source in automation systems. Not only does the operator have random access to thousands of individual songs, but also all the CDs are secured under lock and key.

As noted earlier, regardless of the type of CD player utilized, some compact discs seem to display an overly bright, metalic, or even harsh sound—a distinctive "digital sound." The problem actually has nothing to do with the CD player or even the CD process. It occurs in some CDs because the original recording was intended for vinyl playback. Engineers often added equalization to compensate for

FIGURE 4.6 Cart-style CD player. (Coutesy of DENON America, Inc.)

sonic limitations of vinyl recording. Many CDs are mastered with this same equalization applied rather than removed for transfer to the CD medium, thus the bright, or shiny, sound. Material that has been recorded essentially for CD should not display this problem, nor should older material that has been carefully prepared for transfer to CD.

4.6 RECORDABLE CD PLAYERS

The first CD players were merely turntables for prerecorded CDs, but shortly after their development, manufacturers produced a recordable CD. Broadcast companies like Denon, Studer Revox, and Yamaha all offer **CD recorders** (**CD-R**) that work in the WORM (write once, read many) format (see Figure 4.7). One recordable CD consists of a photosensitive dye layer and a gold reflective layer, encased in the normal protective polycarbonate. When heated by a high-power laser, the dye layer creates bumps (and pits) that can be read like a regular CD. These recorded CDs can be played back on any standard CD player. The biggest drawback to the WORM design is the inability to record over and over; if you make a mistake in the recording pro-

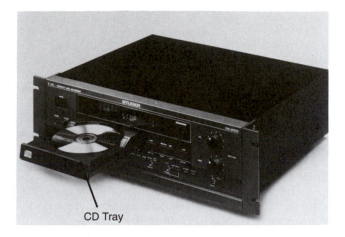

CD Tray

FIGURE 4.7 CD recorder. (Courtesy of Studer Revox America, Inc.)

cess, you have ruined that disc or at least a section of the disc.

Other manufacturers are working to develop a **magneto-optical design** (**MOD**) that would offer a recordable/erasable CD. The recordable CD surface is coated with a magnetic alloy layer that, when heated during the recording process, can be magnetized to change its polarity. During playback, the laser light is reflected differently from opposite-polarity areas, allowing the player to distinguish binary zeros and ones and thus reproduce a digital signal. At this time, this recordable CD is not compatible with current CD players, although it will play on its own recorder, and that recorder will also play standard CDs.

Other approaches to designing a CD recorder are being investigated, but total compatibility with existing CD players remains one of the biggest stumbling blocks to their development. With the intense competition of other digital recording media (such as the MD and DAT, which will be discussed in Chapter Five and Chapter Ten), it is possible that the CD recorder won't be as successful as once envisioned. Still, some stations record jingles, commercials, custom sound effects, and other production elements on CD. In addition, many sound effect and production music libraries that used to be available on record are now produced on compact disc.

4.7 ADVANTAGES OF THE CD PLAYER

The compact disc was the first piece of digital equipment to be embraced in radio production and broadcast work. The CD player offers all the advantages of any piece of digital equipment—greater frequency response, better signal-to-noise ratio, improved dynamic range, and almost no distortion. In addition, there is no physical contact between the player and the actual CD. Unlike the stylus, which rides the grooves of a record, the CD is read by a laser light, and there is no wear or degradation of the CD.

The CD format also offers the convenience of random access to the material stored on the disc. Unlike tape-based formats that require winding through the tape to move from one point to another, CD players can instantly move from one track on the disc to another track anywhere on the disc in a second or two. Another plus of the CD player, from a production viewpoint, is its ability to "cue to music." Whether using a cue wheel or an auto-cueing feature, most CD players allow the announcer to find the exact beginning of the music and start the CD instantly at that point. CD timing information (elapsed/remaining time, track length) is helpful for production work, and the ability to select and play segments of music or to automatically fade-in at any point of the music can be a creative production tool.

The ease of cueing and the consistently high quality obtained makes the use of CDs and the CD player an important part of the production process in the modern studio.

4.8 VINYL RECORDS

Vinyl records are thin plastic discs either 7 inches (the 45 RPM single) or 12 inches (the 33⅓ RPM LP) in diameter. Some broadcast "singles" are produced on the 12-inch disc but are recorded at the 45 RPM speed. **RPM** stands for revolutions per minute, and is just that: the number of times the record goes around in a minute. Beginning broadcasters have often played records at the wrong speed because they automatically assumed the 12-inch disc was a 33⅓ RPM record.

Records start out as a blob of plastic, which is heated and stamped between two metal plates. The master plates are essentially a negative of the record that, when stamped into the plastic, produces the spiral groove (which goes from the edge to the center hole) of the record. The music, or audio signal, is represented in an analog fashion by tiny wiggles in the walls (or sides) of the record groove. The stylus follows the undulations in the record groove to produce an electrical current in the cartridge and ultimately reproduce the original audio signal.

The 45 was originally designed to hold about 4½ minutes of music, and the LP could contain about 30 minutes of music on each side. By spacing the grooves closer, these times have been expanded slightly.

4.9 BROADCAST TURNTABLE FUNCTIONS

In a radio production facility, the **turntable** has two functions: to spin a record at the precise speed at which it was recorded and to convert the variations in the grooves of the record to electrical energy (utilizing the tone arm).

Unlike some home phonographs, a turntable is not a record player. A **record player** is a self-contained unit that not only spins the record and picks up the signal but also amplifies the signal through a speaker. A turntable designed for professional use requires some other means of amplification. Usually the signal is first sent into a preamplifier and then into the audio console. The function of the preamplifier is to increase the level of the signal produced by the cartridge/stylus before sending it to the audio console for further amplification.

Professional-grade turntables are different from consumer-grade turntables in that the former are:

a. heavy-duty (designed for long hours of continuous use)
b. extremely accurate (maintaining precise speeds)
c. capable of quick speed buildup
d. housed in a sturdy base or cabinet
e. capable of being cued (the record or platter can be turned backward)

4.10 BASIC TURNTABLE PARTS

The basic parts of the broadcast-quality turntable are shown in Figure 4.8 and include a platter (a heavy metal plate about 12 inches across covered by a felt or rubber top), a motor to turn the platter, a tone arm, a cartridge/stylus, and a preamplifier. Additionally, most production turntables have an on/off switch and speed selector switch that work in conjunction with the motor. Some broadcast turntables offer the option of an equalizer/filter switch or a pitch control.

The on/off switch and the **speed selector switch** work with the motor to rotate the **platter** at a precise speed. The on/off switch controls power to the motor. In most broadcast facilities, this motor is designed to run for long periods of time. This switch may be left in the on position as long as the control board is in use. Most broadcast turntables use only the 45 RPM and 33⅓ RPM speeds, though some older turntables may include a 78 RPM selection. If the turntable has a pitch control, it just varies the normal speed, usually plus or minus 5–10 percent. In addition to these speed selections, the speed selector switch may have a neutral position. Turntables should be left in neutral when not in use. The neutral position also allows the operator to move the turntable in either direction by hand for cueing.

The basic types of drive systems used in professional turntables include the **idler-wheel turntable**, **belt-drive turntable**, and **direct-drive turntable**. In a direct-drive turntable, the platter sits on top of the motor. In fact, the shaft of the motor is the spindle of the turntable. Precise speeds are controlled electronically rather than mechanically as they are in the idler-wheel and belt-drive systems. The majority of broadcast turntables use a direct-drive system.

Some professional-quality turntables include an **equalizer/filter switch**. Filters and equalizer switches can compensate for poor recordings or records that have been scratched. These scratches usually appear in the high frequencies of the sound. Using the equalizer on a setting for poor records will chop off the high frequencies of the record, eliminating some of the scratch. Not all scratches can be eliminated easily. For example, some appear not only in the higher frequencies, but also the upper midrange. Filtering or equalizing so much of the audio signal can greatly change how the record sounds, perhaps to the point of making it unusable.

4.11 TONE ARM/CARTRIDGE/STYLUS ASSEMBLY

A **tone arm** is usually a metal tube attached to a pivot assembly near the back of the turntable. At the front is a **headshell**, which allows installation of the cartridge/stylus

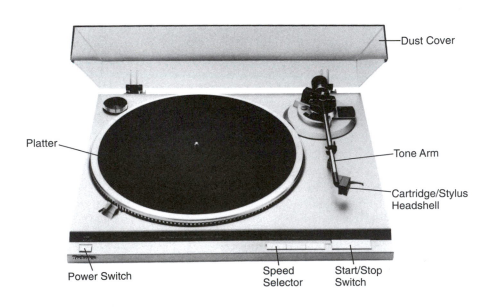

Platter

Dust Cover

Tone Arm

Cartridge/Stylus
Headshell

Power Switch

Speed
Selector

Start/Stop
Switch

FIGURE 4.8   Basic turntable parts. (Courtesy of Panasonic Company)

assembly. The function of the tone arm is to house the cartridge and stylus and allow them to move freely across the record as it is played. This is a very sensitive piece of equipment and should be handled carefully. Some tone arms have small handles next to the cartridge that should be used to pick up the arm and place it on the record. This needs to be done carefully to avoid damage to the cartridge/stylus and to avoid changing the weight and angle of the tone arm.

The weight of a tone arm is called **tracking force**. It's usually under 2 grams, and it can be adjusted by counterbalancing weights at the rear end of the tone arm. This adjustment requires a tracking force gauge and is best left to the station engineer. Improper weight results in damage to the record or poor tracking ability. Never place additional weight (such as a dime or paper clip) on the tone arm to increase its tracking force.

The cartridge/stylus assembly is the working end of the tone arm that actually picks up the signal from the record (see Figure 4.9). The **cartridge** receives the minute vibrations from the **stylus** and converts them into variations in voltage. It acts similarly to the sound-generating element in a microphone (another transducer) and sends the signals to the control board. The stylus is a very small, highly compliant strip of metal. The end that touches the record groove is made of a hard material, usually diamond. The other end is a tiny magnet situated between two coils of wire wrapped around iron cores mounted in the cartridge. As the record grooves vibrate the stylus, the magnet's motion generates a small voltage in the coil windings in the cartridge. This electrical signal is then sent to the preamplifier. This common cartridge design is called a **moving-magnet**. Another type of cartridge is the **moving-coil**, in which small wire coils are attached to the stylus and situated between fixed magnetic structures in the cartridge. Again, movement of the coils within a magnetic field produces a signal that can be sent to the turntable preamplifier. Moving-coil cartridges produce a rather low-level output that might need additional amplification and is subject to hum, so you're more likely to find a moving-magnet cartridge on your production-room turntable.

Styli are either spherical or elliptical; spherical styli are preferred for broadcast use because they allow backtrack-

ing of the record (for cueing up) with a minimum amount of damage to the vinyl grooves. The stylus should not be touched with the fingers. If it's necessary to remove dust from the stylus, do this by blowing lightly on it, or use a special fine-hair stylus brush. Brush lightly from back to front on the stylus.

4.12 CARE OF RECORDS

Dust is probably the biggest problem for records and turntables. Dust in record grooves causes undue record and stylus wear. Ultimately, dust develops permanent pops and scratches on the record. The static electricity produced by playing a record compounds the problem by attracting more dust. Use a good-quality record cleaner before playing records to help minimize the dust problem. Unlike CDs, records can be wiped with a cloth and record cleaning fluid (to remove dust or fingerprints), using a circular motion following the record grooves.

Records should be handled by the edges to avoid getting fingerprints on the grooved surface, and, unless they're being played, keep records in their paper/plastic inner sleeve and cardboard jacket to keep dust off them.

Both LPs and 45s should be stored in a vertical position to prevent them from warping. Most broadcast record storage cabinets are designed to store records in this fashion. It also makes it a lot easier to remove the one record you want from a stack of records.

4.13 CUEING RECORDS

In most broadcast and production situations, you'll want the sound to begin immediately when playing a record. Broadcasters cue the record to avoid any silence (**dead air**) between the start of the record and the first sound heard. To cue a record:

a. Gently place the stylus in the outer groove of the record
b. Put the gear selector switch in neutral
c. Rotate turntable platter clockwise until the first sound is heard
d. Backtrack the turntable platter enough to avoid wowing the record—usually one-eighth turn for 33⅓ RPM records and one-quarter turn for 45 RPM records
e. Put the speed selector switch in the proper speed
f. Turn the on/off switch to on just before you want actual sound to begin

Another way to cue a record is known as **slip cueing**:

a. Cue as above in steps *a* through *e*
b. Hold the edge of the record with a finger with enough force to keep it from spinning when the on/off switch is turned to *on* (the turntable platter will be spinning below the record)
c. Release the record when you want the actual sound to begin

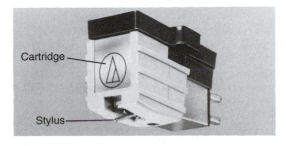

FIGURE 4.9 Cartridge/stylus assembly. (Courtesy of Audio-Technica)

You can only attempt to slip cue if the turntable has a felt-type mat. The typical rubber mat on most home turntables and some professional ones doesn't allow the platter to continue to spin as you hold the record edge.

Slip cueing allows tighter cueing than conventional cueing, but it takes some practice to become proficient at it. The slip-cue technique is often used in production to lift a specific phrase or portion of music off a disk so that it starts cleanly (no wow) and at a precise time.

Both methods for cueing records, but especially slip cueing, cause some deterioration of the outer grooves of the record as the stylus moves back and forth. After a period of time, most broadcast records develop **cue burns**—the name broadcasters give to the damaged outer grooves of the record.

4.14 "WOW" AND "DEAD ROLLING"

A common problem with turntables is improper playback speed, which results in a distorted reproduction of the original sound signal. **Wow** refers to changes in pitch caused by slow, regular variations in the playback speed, and the term is somewhat descriptive of the actual sound heard. It's commonly caused by not backtracking far enough when cueing a record, so the record wows as it builds up to speed when the turntable starts to play. Wow can also be heard because of a record defect (warp or off-center hole) or a turntable defect (worn motor bearings or idler-wheel dip). In any case, wow should be avoided in radio production and broadcast work.

In production work it's frequently important to end a piece of music at an exact time. For example, at the end of a program you might want the music and closing narration to end at the same time, or a radio show's musical theme to end right at the top of the hour so that a newscast can begin. Regardless of the situation, you won't often find music that fits exactly the time that you have to fill. One method of dealing with this is the **dead roll**. To dead roll a record means to begin playing the record with the volume turned down. If you had a record 4:30 minutes long but needed to fill 3:00 before the top of the hour, you would dead roll the record at 55:30, and fade in the music at 57:00. In other words, the first 1:30 of the music would not be heard, but the music would end exactly as desired at the top of the hour. Obviously you usually dead roll music that's instrumental, so it isn't noticeable that you have cut out part of the song.

4.15 CONCLUSIONS

Turntables were a primary workhorse in most radio stations for years. With the advent of compact discs and the quality and convenience they offer, the turntable's use continues to diminish. However, being comfortable working with both the CD player and the broadcast turntable should come fairly easily because most radio production people have already had experience with their home stereo systems. Remember the differences between home stereo systems and broadcast-quality CD players and turntables, and observe good production techniques to improve your equipment operation.

Self-Study

■ QUESTIONS

1. In the digital recording process, an electromagnetic representation of the sound wave of the original sound source is stored on magnetic tape.
   a) true
   b) false

2. Digital recordings can be copied over and over with no measurable loss of sound quality because _____.
   a) the sound signal is a continuously variable electrical signal, the shape of which is defined by the shape of the sound wave produced
   b) the original sound source is sampled over 44,000 times per second
   c) it is binary data that is recorded on the tape, and this can be accurately copied from tape to tape
   d) the laser beam is focused directly on the data track

3. One reason for using a professional-quality CD player rather than a consumer model in the radio production studio is _____.
   a) the professional-quality CD player uses a higher-powered laser to read the CD than a consumer model

b) the professional-quality CD player offers a better signal-to-noise ratio and less distortion than the consumer model

c) the professional-quality CD player offers a greater frequency response and better dynamic range than the consumer model

d) the professional-quality CD player is built for heavy-duty, continuous operation, and the consumer model isn't

4. The lacquer coating on a compact disc makes it virtually indestructible in normal broadcast use.
    a) true
    b) false

5. The CD/CD player format is an improvement over the vinyl record/turntable format because _____.
    a) CDs reproduce analog sound more accurately than records
    b) CD players are more durable than turntables
    c) CD players don't wear out discs as turntables wear out records
    d) CD players have better cartridges than records

6. A multiplay CD player can _____.
    a) hold up to 200 CDs
    b) record
    c) utilize a higher sampling rate than a regular CD player
    d) play two CDs at the same time

7. The standard CD sampling rate is _____ times per second.
    a) 32,000
    b) 44,100
    c) 48,000
    d) 88,200

8. A CD recorder that can record a blank CD only once is using a _____ format.
    a) MOD
    b) CD-R
    c) WORM
    d) DVD

9. Both CDs and records can be easily cleaned with a cloth wiped in a circular motion over the playing surface of the disc.
    a) true
    b) false

10. Which of the following is a function of a turntable?
    a) spinning a record at the proper speed
    b) allowing the cartridge/stylus to move freely across the record
    c) amplifying the signal
    d) none of the above

11. Which of the following picks up a signal from a record?
    a) equalizer
    b) tone arm
    c) speed selector switch
    d) cartridge/stylus

12. The part of the turntable that changes vibrations into variations in voltage is the _____.
    a) cartridge
    b) stylus
    c) diamond
    d) tone arm

13. All 12-inch vinyl records are played back at the $33\frac{1}{3}$ RPM speed.
    a) true
    b) false

14. Which stage of the digital recording process breaks down the analog signal into discrete values or levels?
    a) filtering
    b) sampling
    c) quantizing
    d) coding

15. On the speed selector switch, the neutral position is used for _____.
    a) 45 RPM records
    b) $33\frac{1}{3}$ RPM records
    c) pitch control
    d) cueing

16. The tone arm is used to _____.
    a) pick up sound
    b) contain the instruments that pick up sound
    c) act as a weight
    d) prevent wow

17. A purpose of a turntable equalizer is to _____.
    a) get rid of scratchy noises
    b) compensate for room noises
    c) pick up vibrations
    d) prevent wow

18. Which turntable drive system maintains precise speeds electronically rather than mechanically?
    a) idler-wheel drive
    b) belt-drive
    c) laser-drive
    d) direct-drive

19. To begin a record (or any sound source) with the volume control turned down is to _____ the record.
    a) cue
    b) wow
    c) dead air
    d) dead roll

20. As a final test on CDs and turntables, match the items in the top list (1, 2, 3 . . .) with the choices in the bottom list (d, f, o . . .), and then select the correct set of answers from the sequences shown in a, b, c, or d below.

    1. _____ analog signal
    2. _____ laser diode
    3. _____ tone arm
    4. _____ cue wheel
    5. _____ digital signal
    6. _____ stylus
    7. _____ motor
    8. _____ equalizer/filter switch
    9. _____ reflective coating
    10. _____ tray
    11. _____ prism system
    12. _____ speed selector switch
    13. _____ cartridge

 d. directs a laser light beam toward the CD
 f. allows for free movement across a record
 o. often has a diamond point
 s. a signal whose shape is defined by the shape of the sound wave produced
 r. allows a CD to be "rocked"
m. makes the turntable platter rotate
 b. a discrete signal composed in a binary manner
 l. makes it possible for a laser to read data on a CD
 p. where a CD is placed in a CD player
 g. generates a laser light beam
 u. used to switch between records recorded at $33\frac{1}{3}$ RPM and at 45 RPM
 v. converts vibrations to variations in voltage
 n. gets rid of scratchy record noises

 a) 1.b 2.g 3.f 4.r 5.s 6.o 7.m 8.n 9.l 10.p 11.d 12.u 13.v
 b) 1.s 2.g 3.f 4.r 5.b 6.o 7.m 8.n 9.l 10.p 11.d 12.u 13.v
 c) 1.s 2.o 3.l 4.v 5.b 6.d 7.p 8.n 9.f 10.m 11.u 12.r 13.g
 d) 1.b 2.d 3.f 4.r 5.s 6.v 7.m 8.n 9.l 10.p 11.g 12.u 13.o

■ ANSWERS

If You Answered A:
 1a. No. This statement describes analog recording, not digital recording. (Reread 4.2.)
 2a. Wrong. This refers to an analog signal. (Reread 4.2.)
 3a. No. Both professional and consumer-model CD players use a similar laser system. (Reread 4.5.)
 4a. No. Although the lacquer coating helps protect the CD, it is far from indestructible—fingerprints, dust, and scratches have damaged CDs. (Reread 4.3 and 4.4.)
 5a. Wrong. CDs use digital, not analog technology.(Reread 4.2, 4.5, and 4.7.)
 6a. Right. A multiplay CD player holds many CDs, which could enable a station to be automated.
 7a. No. This is another sampling rate that some digital equipment is capable of utilizing. (Reread 4.2 and 4.3.)
 8a. No. This is a recordable/erasable format. (Reread 4.6.)
 9a. Wrong. CDs should only be wiped in a straight line from the center hole toward the outer rim. (Reread 4.4 and 4.12.)
10a. Correct. This is a function of the turntable.
11a. No. (Reread 4.10 and 4.11.)
12a. Right. The cartridge converts vibrations into voltage.
13a. Wrong. Some 12-inch "singles" are designed to play back at the 45 RPM speed. (Reread 4.8.)
14a. No. Filtering cuts off unwanted high frequencies before sampling begins. (Reread 4.2.)
15a. No. There's a separate position for this. (Reread 4.10.)
16a. No. It isn't actually responsible for making sound. (Reread 4.11).
17a. Yes. The equalizer/filter can help get rid of scratchy noises.
18a. No. The idler-wheel motor-pressure roller-platter drive system is mechanical. (Reread 4.10.)
19a. No. To cue a record is to prepare it for air play, but this doesn't describe it. (Reread 4.13 and 4.14.)
20a. Almost but not quite right. You are confused about analog and digital. (Reread 4.2.)

If You Answered B:
 1b. Correct. This is not digital recording, but rather analog recording.
 2b. No. But you're heading in the right direction. Sampling rate is important for exact reproduction of the original sound, but it isn't really the reason why digital recordings can be copied over and over. (Reread 4.2.)
 3b. Wrong. All CD players have similar S/N and distortion characteristics. (Reread 4.2 and 4.5.)
 4b. Yes. CDs require careful handling even though the lacquer coating helps prevent problems.
 5b. Wrong. Both professional-quality CD players and turntables are durable. (Reread 4.5 and 4.7.)
 6b. Wrong. CD multiplay players are not designed to also record. (Reread 4.5.)
 7b. Correct. This is the standard sampling rate for CDs.
 8b. No. This is a generic term for all CD recorders. (Reread 4.6.)

9b. Correct. While a circular cleaning motion works for vinyl, it can damage a CD.

10b. You're close. The tone arm (which is part of a turntable) houses the cartridge/stylus and allows them to move across the record. (Reread 4.9 and 4.11.)

11b. You're close. The tone arm houses the cartridge/stylus, but it doesn't pick up any signal. (Reread 4.11.)

12b. Wrong. The stylus picks up vibrations. (Reread 4.11.)

13b. Correct. Many 12-inch records play at $33\frac{1}{3}$ RPM, but some are designed to play at 45 RPM.

14b. No. Sampling is when the signal is "sliced" into thousands of individual samples (voltages). (Reread 4.2.)

15b. No. There's a separate position for this. (Reread 4.10.)

16b. Right. It houses the stylus and cartridge, which pick up sound.

17b. No. (Reread 4.10.)

18b. No. The belt drive system of motor-belt-subplatter is mechanical. (Reread 4.10.)

19b. No. Wow is an off-speed problem associated with records, but this doesn't describe it. (Reread 4.13 and 4.14.)

20b. Very good. You have now completed the Self-Study section on CD players and turntables.

If You Answered C:

2c. Yes. Binary data can be accurately recorded over and over, making digital copies sound exactly like the original.

3c. No. All CD players have similar frequency response and dynamic range characteristics. (Reread 4.5.)

5c. Right. Since there is no physical contact between CD and CD player, the CD doesn't wear out as a record's grooves will.

6c. Wrong. Sampling rate has nothing to do with multiplay. (Reread 4.3 and 4.5.)

7c. No. This is another sampling rate that some digital equipment is capable of utilizing. (Reread 4.2 and 4.3.)

8c. Yes. This is a write once, read many format.

10c. No. A turntable itself does not amplify. Amplification is done through the preamplifier and the audio console or control board. (Reread 4.9 and 4.10.)

11c. No. (Reread 4.10 and 4.11.)

12c. Wrong. This is part of the stylus. (Reread 4.11.)

14c. Yes. This is correct.

15c. No. A separate control, not the speed selector switch, is used to change pitch. (Reread 4.10.)

16c. No. That is not its function. (Reread 4.11.)

17c. No. That's the cartridge. (Reread 4.10 and 4.11.)

18c. Wrong. There is no such thing. (Reread 4.10.)

19c. No. You're close because dead air equals silence, but this doesn't really describe it. (Reread 4.13 and 4.14.)

20c. No. You have confused CD players and turntables. (Reread the entire chapter)

If You Answered D:

2d. No. While the use of a laser is an important part of CD digital technology, it is not the reason. (Reread 4.2 and 4.5.)

3d. Yes. Most consumer-model CD players can't stand up long to the daily use of the broadcast facility.

5d. No. CD players do not contain a cartridge. (Reread 4.5, 4.7, and 4.11.)

6d. No. (Reread 4.5.)

7d. Wrong. This would be a two-times-over sampling rate. (Reread 4.2 and 4.3.)

8d. No. There is a correct answer. (Reread 4.6.)

10d. No. There is a function of the turntable listed. (Reread 4.9.)

11d. Right. The cartridge and stylus are the working end of the turntable tone arm.

12d. No. (Reread 4.11.)

14d. No. Coding is when a series of binary digits are assigned to each individual sample. (Reread 4.2.)

15d. Right. In the neutral position, the turntable moves freely.

16d. No. You're way off base. (Reread 4.11 and 4.14.)

17d. No. (Reread 4.10 and 4.14.)

18d. Yes. The direct-drive system's precise speeds are controlled electronically.

19d. Yes. This describes a dead roll.

20d. No. You made a few mistakes. (Reread 4.2, 4.5, and 4.11.)

Projects

■ PROJECT 1

Play and record several compact disc selections.

Purpose

To familiarize you with the operation of CD players.

Advice, Cautions, and Background

1. You will have to do this in conjunction with a microphone, audio board, and tape recorder, so you may need help from your instructor to operate the pieces of equipment that you haven't yet learned.
2. Different brands of CD players have slightly different features, so you will need to learn the particular characteristics of your player.
3. You can play several selections from one CD or use several CDs, whichever you prefer.

How to Do the Project

1. Make sure your CD is connected so that it can be faded out and will record onto an audio tape recorder.
2. Make sure a microphone is available so that you can announce the title of the selections you choose.
3. Examine the CD player and practice with it so that you can cue, play, and pause it.
4. When you feel familiar with the player and have decided on three selections to play, start the tape recorder.
5. Complete the project by doing the following:
    a. Announce the name of the first musical selection, bring it in, and then fade it out after about 30 seconds.
    b. Announce the name of the second selection, bring it in and fade it out after about 30 seconds. (If you have two CD players, you can cue the second one while the first is playing. If not, you'll need to cue a second selection while you are introducing it. This is not particularly difficult because most CD players enable you to cue easily.)
    c. In the same manner, announce the name of the third selection, bring it in, and then fade it out.
6. Label the tape with your name and CD Recording.
7. Turn in the tape to your instructor to receive credit for the project.

■ PROJECT 2

Cue several records, using both the regular cueing method and the slip cue method.

Purpose

To enable you to feel comfortable working with turntables.

Advice, Cautions, and Background

1. If you're not sure of what you're doing, ask the instructor for assistance. Don't take the chance of ruining the equipment, especially the cartridge/stylus assembly.
2. If your turntables have rubber mats on the platters, you may not be able to do the slip cue method.
3. Remember to record your work on audio tape.
4. You'll be judged on proper tightness of record cueing (avoid dead air and wows).

How to Do the Project

1. Familiarize yourself with the operation of the turntables in your production studio. If you have questions, ask the instructor.
2. Cue a record using the standard cueing techniques:
   a. After putting a record on the turntable platter, gently place the stylus in the outer groove of the record.
   b. Put the speed selector switch in neutral.
   c. Rotate the turntable platter clockwise until the first sound is heard.
   d. Backtrack the turntable platter (rotate it counterclockwise) about one-eighth to one-quarter of a turn.
   e. Put the speed selector switch in the proper speed.
3. Make sure your audio tape is recording.
4. Announce the record title and artist in an ad-lib manner.
5. Turn the on/off switch to "on" just before you want the actual sound to begin.
6. Fade out the record after 20 to 30 seconds.
7. Repeat the above steps for at least two more records.
8. Cue a record using the slip cue method:
   a. After putting the record on the turntable platter, gently place the stylus in the outer groove of the record.
   b. Put the speed selector switch in the proper speed.
   c. Turn the on/off switch to on.
   d. When the first sound is heard, hold the edge of the record with your finger, using enough force to keep it from spinning (the platter will continue to rotate).
   e. Backtrack the record about one-eighth to one-quarter of a turn, and continue to hold the edge of the record.
9. Make sure your audio tape is recording.
10. Announce the record title and artist in an ad-lib manner.
11. Gently release the edge of the record just before you want the actual sound to begin.
12. Fade out the record after 20 to 30 seconds.
13. Repeat the above steps for at least two more records.
14. Label the tape with your name and RECORD CUEING.
15. Turn in the tape to your instructor to receive credit for the project.

CHAPTER FIVE

Audio Tape Recorders

Information

5.1 INTRODUCTION

There are three basic kinds of **audio tape recorders** found in the radio production facility: reel-to-reel recorders, cartridge recorders, and cassette recorders. They are all similar in that they store electrical impulses that can be changed back into sound, but their construction and mode of operation differ. Originally they all recorded in an analog fashion, but now several different forms of digital recorders exist. **Open reel** and **closed reel** systems are two terms often used to categorize audio tape recorders. Reel-to-reel recorders are considered open reel systems because you physically handle the tape when you are threading it on the reels of the recorder. Closed reel systems have small reels of tape enclosed in a plastic casing that is designed for a specific recorder, such as the cassette or cartridge. Closed reel systems are more convenient to use, and this is one of the reasons they gained popularity in broadcast use.

5.2 REEL-TO-REEL RECORDERS

The early parts of this chapter deal with the reel-to-reel recorder, although many of the general concepts are equally true for the cartridge and cassette recorders. Often the term "recorder" is used, even though, as we will see, some of the machines can only play back. Developed in the mid-1930s, the reel-to-reel audio tape recorder has been the workhorse of the production room. The basic "parts" of a typical reel-to-reel recorder consist of the tape recorder heads, the tape transport, and the electronics of the recorder.

5.3 HEAD ARRANGEMENT

Tape recorders are devices that rearrange particles on magnetic tape so that sound impulses can be stored on the tape and played back later. This rearranging of particles is done by **heads**. You'll learn more about audio tape in the next chapter, but for now it's useful to understand that the magnetic layer of tape consists of metallic particles. When a tape is unrecorded, the magnetic particles are not aligned and are on the tape in a random pattern (see Figure 5.1).

Usually, professional-quality recorders have three heads: one to erase, one to record, and one to play. The erase head is always before the record head so that old material can be erased and new material recorded at the same time. It is erased an instant before the new material is recorded on the audio tape. With the play head behind the record head, it's possible to monitor what you've just recorded. When the machine is just in play, the erase and record heads are disengaged. The easiest way to remember the arrangement of heads in an audio tape recorder is to remember: "Every recorder plays," or "ERP"—erase, record, play.

If the recorder appears to have a single head, it may be because the erase head has been combined into the same case with a record/play head. However, some playback-only machines truly have just a single play head. Less expensive recorders (and many home recorders) have only two heads: one for erase and another for both play and record. Regardless of the possible head arrangements, ERP is the most common one for audio tape recorders found in the radio production room.

The three-head configuration of professional recorders offers several advantages. Each head can be optimized in design and construction for the specific function that it serves—erasing, recording, or playing back audio tape. Individual heads are easier to align and maintain. The three-head design also allows the operator to monitor the sound quality directly off the tape during the recording process.

5.4 SEL SYNC

One feature often found on audio tape recorders is **sel sync** (selective synchronization), also known as sound-on-sound. Because of the ERP head arrangement, you can't record one track in synchronization with a previously recorded track. For example, if you record one voice and want to record another voice on the same tape, you run into this problem: As the previously recorded voice is playing, the sound signal is coming from the play head, but the second voice is recording at the record head, and because of the distance between these two heads, you hear the previously recorded material a split second before you can record the second voice. To overcome this problem and allow

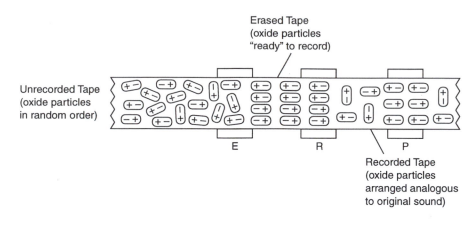

FIGURE 5.1 Head arrangement and tape recording.

this type of recording, the sel sync feature makes the record head also act as the play head. Now you're hearing the previously recorded material at the same time as you're recording the new material, so there is no time difference between them, and you can easily synchronize the two recordings (see Figure 5.2). This can be an important production tool, especially when you're doing multivoice spots.

5.5 TAPE RECORDER FUNCTIONS—ERASE, RECORD, AND PLAY

The erase, record, and play heads are very similar. Each head consists of a laminated metal core that is wound with an extremely fine wire coil at one end. At the other end of the core is a gap between the two magnet poles of the head (see Figure 5.3). Tape recorder heads are merely small electromagnets; an electrical current through the coil creates a fluctuating magnetic field at the head gap.

When recording, the audio tape is pulled across the head gap at a right angle. The sound signal is delivered to the re-

cord head from the record amplifier in the tape recorder and is transformed from an electrical signal to a magnetic signal. This magnetic signal jumps the head gap and magnetizes the iron oxide layer of the audio tape (remember, the tape was just erased), passing by in a pattern analogous to the original sound signal (refer to Figure 5.1).

Unfortunately, because the oxide coating on audio tape responds to magnetization differently at low and high input signal levels, it won't record an exact analog of the original sound source at normal levels and could be severely distorted at lower levels. To compensate for this, a bias current is added during the recording process. **Bias** is an inaudible, high-frequency (at least 100,000 Hz) tone that, when mixed with the audio signal, raises the overall level to a point where the audio tape records more evenly.

The addition of bias improves the frequency response of the recording and provides a distortion-free signal. If the recording sounds too bright and fuzzy, there may be too little bias used; too much bias can result in the loss of the audio recording's high frequencies. Proper bias settings are accomplished by the recorder manufacturer or station engi-

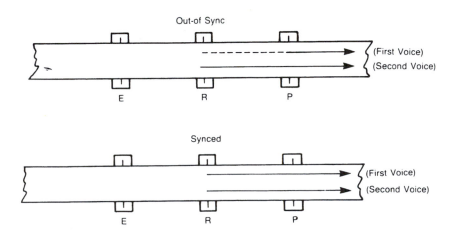

FIGURE 5.2 Sel sync recording.

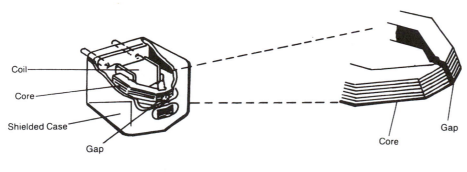

Internal View

FIGURE 5.3 Audio tape recorder head.

neer and are usually set for one particular brand of audio tape. As a radio production person, avoid using a hodge-podge of different audio tapes, and stick to the brand that your tape recorders have been set to, for best recording results.

During play, the recording process is reversed. A recorded (magnetized) tape is drawn across the gap of the play head. The magnetic field of the tape (at the gap) passes into the core and then into the coil, creating an electrical current. This current is sent to a play amplifier in the recorder and is an exact reproduction of the original sound signal.

Audio tape reproduces low and high frequencies at a softer level than the middle frequencies, and for that reason **equalization** is often used. Equalization affects the amount of amplification that is given to the highs and lows of the sound signal. Generally, the highs are boosted during recording and slightly decreased during playback; the low frequencies are increased during playback. The ultimate goal is a flat signal response. Like bias, equalization adjustments are often handled by the engineer, although some tape machines do have controls that can be adjusted by the radio production person.

The erase head is on during the recording process. A magnetic field is produced at the erase head gap that is so powerful, it demagnetizes the audio tape as it passes by. During the erase process, the random pattern of metallic particles on the unrecorded audio tape are arranged in a pattern that makes them ready to be recorded again (see Figure 5.1).

5.6 TAPE RECORDER SPEEDS

Most reel-to-reel tape recorders found in the production studio can record at different speeds, ranging from $^{15}/_{16}$ IPS (inches per second) to 30 IPS. The most common are $3\frac{3}{4}$ IPS, $7\frac{1}{2}$ IPS, and 15 IPS. Of course, you must play back the tape at the same speed at which it was recorded. If you don't, it will have a speeded up "Donald Duck" sound or a slow "groggy" sound.

A tape machine operating at $7\frac{1}{2}$ IPS means $7\frac{1}{2}$ inches of recording tape go past the head each second. The more tape that goes past the head, the better the recording because greater frequency response can be put on the tape with a better signal-to-noise ratio. Most radio production work is done at $7\frac{1}{2}$ IPS or 15 IPS. You'll find that if you are editing audio tape, the faster the tape recorder speed, the easier it is to edit. (Obviously, the pauses between words, etc., will be longer at the faster speed.) A tape recorder speed of $3\frac{3}{4}$ IPS is usually acceptable for recorded material that is voice only. To get high-quality music recordings, avoid the slower speeds (most recording studios use tape recorders that record at 30 IPS and higher). The advantage of the slower speed is that you can get more material on the tape. Most modern tape recorders can be set to operate at a number of different speeds.

5.7 TAPE COUNTERS/TIMERS

Most audio tape recorders have a built-in timer or tape counter. Modern machines have a digital timer that gives an accurate minutes and seconds count. Some machines have a mechanical counter that counts up from "0000," but the numbers have no relationship to time and are merely a gauge as to where you are on the tape. Most timers and counters also have a "zero set" that allows you to rewind to an exact location. For example, at the beginning of a voice track you're recording, you set the timer (or counter) to "00:00" and enable the zero set button. When you complete the recording, just push the rewind button on the recorder, and it will rewind to the point where you set the "00:00." Timers are also useful to get a timing of a longer radio program. You don't have to listen to the program in real time since most timers work in the fast forward mode of the recorder. An accurate timing of a half-hour radio show may only take a few minutes in fast forward; just reset the timer to "00:00" at the first sound at the beginning of the tape, fast forward to the last sound at the end of the tape, and look at the timer reading.

5.8 TAPE RECORDER TRANSPORT

Figure 5.4 shows the face of a typical reel-to-reel recorder that is found in the production studio. The labeled parts in this diagram are referred to as parts of the **tape transport**. As the name suggests, the tape transport is that part of the recorder that's involved with the actual motion of the audio tape as it passes the tape recorder heads. Starting with the tape reels, since the audio tape threads on a recorder from the left to the right, the left reel is the **supply reel**, or the **feed reel** (the reel that has audio tape on it as you begin to use the recorder). The right reel is the **take-up reel**, which starts out empty. Behind each reel (inside the tape recorder) are motors that help drive the tape from one reel to the other.

The three standard reel sizes used in radio production facilities—5 inch, 7 inch, and 10½ inch—are shown in Figure 5.5. The 10½-inch reel shown in the diagram has an NAB hub (center) that requires an adapter when used with most recorders. It's important that both the feed reel and the take-up reel be the same size for keeping the audio tape at the proper tension. There is often a reel-size switch on the recorder that the operator should set properly. The small reel setting generally refers to 7 inch or smaller reels, and the large reel setting is for the 10½ inch reels.

The audio tape is kept in line with the tape heads by various tape guides and tension arms. The **tape guides** are usually just stationary pins that provide a track or groove the width of the audio tape. The **tension arms** are generally movable. As the audio tape threads through them, they provide some spring, or tension, against the tape. There is often a tension arm on each side of the tape heads. One of these tension arms is usually an **idler arm**. In addition to providing proper tension on the audio tape, if the tape breaks, this arm drops down into an off position, and the reel-to-reel recorder stops running, so the tape doesn't spill off the feed reel and all over the studio floor. The idler arm also keeps the recorder transport in neutral even though it is powered on, which helps prolong the life span of the machine.

The heart of the tape transport is the capstan and pinch roller. Normally located just to the right of the tape heads, the **capstan** is a metal shaft, and the **pinch roller** is a rubber wheel. The audio tape passes between these two components. When the recorder is running, the pinch roller holds the tape against the revolving capstan. The capstan controls the speed of the tape as it passes the heads.

The final components of the tape transport are the actual controls of the audio tape recorder. Most recorders have an on/off button, rewind and fast forward buttons, a play button, a stop button, a pause button, a cue button, and a record button. Most of these functions are obvious and

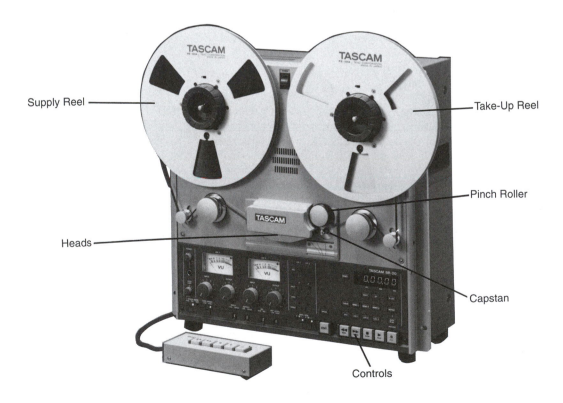

FIGURE 5.4 Audio tape recorder transport. (Courtesy of TASCAM Division, TEAC America, Inc.)

FIGURE 5.5 Reel-to-reel reel sizes. (Courtesy of Ampex Recording Media Corporation)

merely control the direction and speed of the audio tape through the transport. Rewind moves the audio tape from the take-up reel to the feed reel, and fast forward moves it in the other direction. The cue button allows the tape to stay in contact with the tape heads during rewind and fast forward, so you can aurally find a certain spot on the tape. Usually the tape is lifted away from the heads in the rewind or fast forward position to save wear and tear on the heads. Some tape recorders require the operator to depress both play and record to put the machine into the record mode, but some will go into record when just the record button is pressed.

5.9 TAPE RECORDER ELECTRONICS

The other half of the face of the audio tape recorder involves the electronics of the recorder (see Figure 5.6). These controls include record-level and play-level pots, VU meters, and a source/tape switch. The **record-level pot** (pots, if stereo) adjusts the volume or level of the incoming sound signal. If the signal is being fed from an audio console, there is a pot, or channel, on the console that also controls the incoming volume. The best recording procedure is to make sure the level indicated on the audio board VU is near 100 percent, and then fine-tune the volume with the record-level pot on the audio tape recorder. The **play-level pot** controls the volume of the sound signal as it's being played from the audio tape. Again, there is also a volume control that adjusts the output volume of the recorder as it plays through the audio console. These pots should adjust closely together so that you don't have the tape recorder pot

turned way down and then need to turn the audio console pot for the tape recorder way up to get a good signal level.

The VU meter indicates the signal level, just like the VU meters on the audio console that were presented in an earlier chapter. What signal you see on the audio tape recorder VU meter is dependent on where the **source/tape switch** is set. The source position is sometimes labeled "record" or "input," and the tape position is sometimes labeled "play," "reproduce," or "output." In the source position, the VU meter shows the volume of the incoming signal at the record amplifier, usually just before the bias current is added. If this switch is set at the tape position, the VU meter shows the output level of the reproduced signal at the playback head, so it only shows an indication of level when you are actually recording.

You can use the source/tape switch to make comparisons between the two levels. If the tape recorder is properly calibrated, there should be very little level change as you switch between the source and tape positions. Some machines, however, have an output level control that isn't calibrated with the input level control, or dirty record or playback heads, so that you shouldn't always expect the two levels to be equal.

Also associated with the electronics of the audio tape recorder are the inputs found on the back of the recorder. Most tape recorders have provisions for plugging other electronic equipment, such as mics and CD players, into them. The main concept to understand for this is the concept of **preamplification**. As already discussed in Chapter Two, some electronic devices, such as CD players and tape recorders, have provisions within them for amplifying their

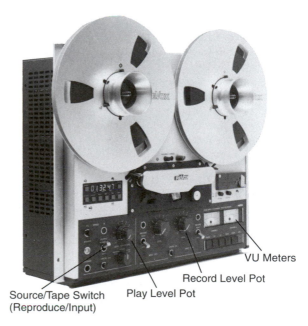

VU Meters
Record Level Pot
Source/Tape Switch   Play Level Pot
(Reproduce/Input)

FIGURE 5.6 Audio tape recorder electronics. (Courtesy of Studer Revox America, Inc.)

electrical impulses. Others, mainly microphones, do not. If a CD player is plugged into a tape recorder, it should be plugged into the **line-level** input because it has already been preamplified in the CD player. In this way the tape recorder won't amplify the sound again, causing distortion. A microphone, on the other hand, should be plugged into a **mic-level** input that allows the tape recorder to amplify the signal.

Most recorders also have a line-out connector so that the sound from the recorder can be fed to other electronic equipment, such as an audio console. Some recorders also have a speaker-out position so that the sound can be transferred to speakers located away from the recorder. Normally in the radio production facility, the audio tape recorders are patched or wired to the other equipment through the audio console, and the operator doesn't have to worry about the various inputs and outputs.

5.10 TRACK CONFIGURATION

Another thing to understand about tape recorders is the recording patterns ("tracks") on tape. Various audio tape recorders record differently by using different portions of the tape. There are four different track configurations that conventional reel-to-reel recorders use. They differ in the number of signals put on the ¼-inch tape and the placement of the signals.

Some recorders are **full-track;** that is, they essentially use the whole ¼-inch space to record one signal. This was the first tape recording method developed and since it's one signal on the whole width of the tape, it's a monophonic recording. Also, full-track recording can occur in only one direction (see Figure 5.7-A). In general, the more tape that is used in the recording process, the higher quality the recording will be, so full-track is a high-quality recording method.

There is also **half-track mono** recording on ¼-inch tape. In this instance, the tape is recorded on twice. The top is recorded on as the tape moves from left to right. Then the tape is flipped over, what was the bottom half becomes the top half, and it's recorded on. The two signals go in opposite directions because the tape has been flipped (see Figure 5.7-B). In no instance is the back of the audio tape recorded on.

To understand recording in both directions (as mentioned above), take a piece of paper and place it on a table or desk between your two hands. Imagine your left hand is the feed reel, the paper is the tape recorder heads, and your right hand is the take-up reel. Draw a line that divides the paper in half, and draw an arrow going from left to right in the upper half. Now imagine the paper moving from your left hand to your right hand. This would be recording in one direction and when you had completed the tape, you would exchange reels. If you turn the paper top to bottom, you will simulate what happens to the audio tape. The top half is now blank, and you can draw another arrow (again going from left to right) to indicate you're recording on the

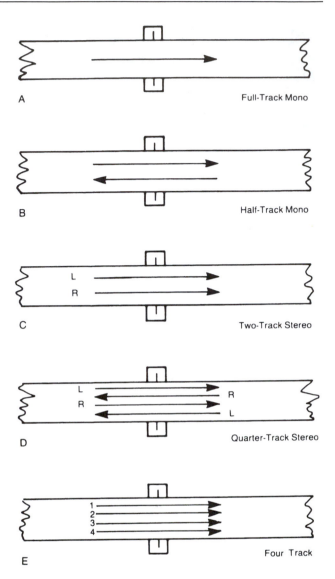

FIGURE 5.7 Reel track configurations: A) Full-Track Mono; B) Half-Track Mono; C) Two-Track Stereo; D) Quarter-Track Stereo; E) Four Track.

other half of the tape. You'll note the arrows are going in opposite directions, and if you turn the tape top to bottom again to put it in its original position, the alignment of the arrows is the same as shown in Figure 5.7-B.

Most recorders used in radio production today are stereo or multitrack. Stereo can use ¼-inch tape, but it requires two tracks for each recording—one for the right channel and one for the left channel. When recording in **two-track stereo** (sometimes called **half-track stereo**), each of the two tracks uses half the tape, and both go in the same direction. One track is for the right channel, and one track is for the left channel (see Figure 5.7-C). Recording can only be done in one direction on this tape because the two tracks use the entire width of the audio tape. This is probably the most common recording method found in the radio production facility.

Many home recorders (and some professional ones) use a **quarter-track stereo** recording method. Tracks one and three (the top one and the next-to-bottom one) are used for the first side, going from left to right. Then the tape is flipped and tracks two and four become one and three and are recorded. The two recordings go in opposite directions (see Figure 5.7-D). The track widths naturally have to be narrower, since four actual tracks have to fit on the ¼-inch audio tape.

Multitrack recorders are becoming more and more common in the radio production facility. For example, a **four-track recorder** uses four separate tracks all going in one direction (see Figure 5.7E). Although a few machines use ¼-inch tape for the four tracks, most multitrack recorders use ½-inch or wider audio tape. Multitrack recorders used in radio are normally four- or eight-track, and the recording studio may employ recorders with sixteen, twenty-four, thirty-two, and more tracks. Obviously, at some point in the production process the multitracked signal is mixed down to stereo or mono for broadcast. Multitrack recording will be discussed further in Chapter Eleven. Four-track recording should not be confused with quarter-track recording. Both use four tracks, but a review of each description will show they are quite different.

5.11 CROSSTALK AND COMPATIBILITY

All the stereo and multitrack recording heads have **guard bands**—spaces between each track that provide small portions of blank tape between each track and at the edges to prevent crosstalk. **Crosstalk** occurs when the signal from one track is picked up simultaneously with the signal from an adjacent track. For example, if the record and playback

heads on a quarter-track recorder were misaligned, you might hear part of side 2 (playing backward) when listening to side 1 because the right channel tracks are adjacent to each other, and the misalignment is causing crosstalk.

Another important consideration is compatibility. Quarter-inch tapes recorded on one machine can play on a different machine. In other words, a tape recorded on a quarter-track stereo machine can be played on a full-track mono recorder. All the tracks will be heard. However, there will be recorded material going both forward and backward, and the resulting sound will be a garble. If the stereo tape were recorded only on one side, then playback on the mono full track would be intelligible, but, of course, it would not be stereo and probably would have some hiss noises from nonrecorded sections. Half-track mono tapes can be played on two-track stereo machines, but they'll be heard only through the left channel. Many two-track stereo recorders found in the production studio have an extra playback head that is quarter-track stereo. A switch allows the operator to select which one will be active at any particular time. Certainly the best production practice would be not to mix track configurations, even though some are compatible to a degree.

5.12 THE CARTRIDGE TAPE RECORDER

Another type of tape recorder is the **cartridge recorder** (see Figure 5.8). The cart machine is favored in broadcast use because it cues up automatically. Many commercials and public service announcements that you hear on the radio are being played from tape cartridge recorders. For some radio stations, even the music they play has been transferred from record or CD to cartridge.

A tape cartridge recorder requires no threading; you simply slip the audio tape cartridge into a machine. Car-

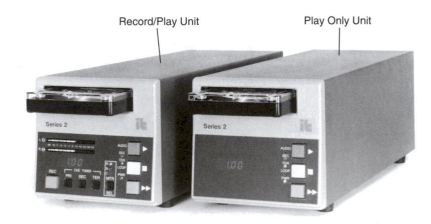

Record/Play Unit   Play Only Unit

FIGURE 5.8 Cartridge recorder and player. (Courtesy of International Tapetronics Corporation)

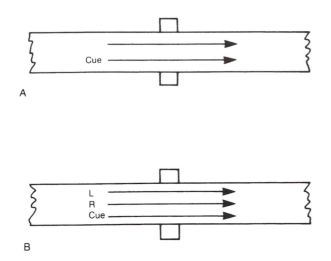

FIGURE 5.9 Cartridge tape configurations: a) mono; b) stereo.

tridge tape is driven by a pinch roller that swings up from the cart recorder through a drive-wheel hole in the plastic cartridge. This happens when the cart is inserted into the machine. The pinch roller presses the tape between itself and the capstan, which is turning. The tape is pulled through in a clockwise motion (the tape travels left to right across the heads). Cart machines operate at 7½ IPS and use ¼-inch tape.

The track configuration of the cartridge recorder is similar to some that you're now familiar with; when cartridge tape is recorded, however, a cart tape head can put a **cue tone** on the tape that can't be heard. This primary cue tone is just in front of the recorded information so that when the machine is running (in play), the special tone signals the machine to stop before it repeats itself. This also allows for putting several different spots on one cartridge and playing one without fear of playing another before stopping the machine. This self-cueing feature is extremely important for radio broadcasting and radio production work. Most cart machines can also put secondary and tertiary cue tones on the cartridge that can be used to indicate the end of a spot, to start another cart machine, and to activate other programming features often found in automated situations.

A mono tape cartridge recorder has two tracks—one for the recorded material and one for the cue tones (see Figure 5.9A). A stereo tape cartridge has three tracks—one for the left and right channel and one for the cue tones (see Figure 5.9B). Mono and stereo tape cartridges are not compatible. Although you may be able to play a cartridge recorded in stereo on a mono machine or vice versa, it's likely that the tracks will be misaligned, and you won't hear all of the audio signal, or there will be problems in picking up the cue tone. Good production practice dictates that mono and stereo cartridges are clearly labeled and not mixed between various tape decks.

Tape carts record in only one direction. Tape cartridge

machines can be either record/play or play-only. The play-only machines still have two heads in them, but one is a dummy that merely keeps the tape tension correct as the audio tape passes the heads.

It's also very important to notice that cartridge tape recorders don't have erase heads in them. Tape carts must be erased using a bulk eraser, or **degausser**. This is merely a strong electromagnet that radiates a magnetic field that erases any audio tape brought near it when it's turned on. More will be said about the degausser in Chapter Eight.

5.13 AUDIO TAPE CARTRIDGES

The audio tape cartridge is constructed as a plastic container with an endless loop of tape. The tape is wound onto the center spool with the end next to the spool sticking up so that when the tape is cut from the master reel, it can be spliced to that end, thus making a continuous loop (see Figure 5.10). The tape pulls from the inside and winds on the outside of the spool. Some carts employ tape guides that the tape goes past and pressure pads to control tension and alignment; others use a design that eliminates the need for pressure pads.

The recording tape used in a cartridge machine is ¼-inch wide, but it's a special type that has a lubricant on both sides rather than just the oxide side. The tape is wound on

FIGURE 5.10 Audio cartridge. (Courtesy of Fidelipac Corporation)

the spool so that the oxide side is facing out, away from the spool. To see if the tape on the cartridge is good, simply look at the oxide side of the tape through the holes on the front side of the cartridge. If the tape is shiny, it's worn (little oxide on it), but if the tape is dull, it probably has quite a bit of oxide left. Often before the tape gives out, the **pressure pads** will. To check the pressure pads on a tape cartridge, just press in on them (go under the tape so that oil from your fingertips doesn't get on the tape); if they tend to stay down and don't spring back, the tape cartridge won't play the tape properly because the pressure pads keep the tape pressed against the tape heads.

5.14 THE CASETTE TAPE RECORDER

The **cassette recorder** has found its way into the broadcast facility mainly because of its portability and ease of use, but also because the modern cassette recorder offers recording quality that often exceeds that of reel-to-reel recorders. Tape machines like the one shown in Figure 5.11 are commonly found in the radio studio. Portable units are also used, primarily in the news area, to record events or conduct interviews. From time to time, however, a cassette recorder is also handy for production work. For example, you might not have a sound effect that you need on record or CD, but you can easily record your own on a portable cassette.

One of the reasons the cassette tape became popular is that it doesn't have to be threaded onto the machine as the reel-to-reel does. Just pop the audio tape cassette in the machine, and you're ready to record. Cassette tape is ⅛-inch wide, so it can never be played back on a ¼-inch reel-to-reel recorder. There are two basic cassette tape recording methods. One is half-track mono, which is like half-track mono

on a ¼-inch tape except that it records the bottom of the tape first (see Figure 5.12A). The other recording method for cassettes is quarter-track stereo. This is not the same as ¼-inch quarter-track stereo (review Figure 5.7D); rather, the bottom two tracks are used for one side going in a left-to-right direction. When the tape is turned over, the top two tracks are the bottom two, so they're recorded going in the opposite direction (see Figure 5.12B). Cassettes were designed this way so that mono and stereo cassettes would be compatible.

There are numerous cassette recorders available. Many are only suitable for home use and not the broadcast studio. Most portable cassette recorders have a built-in microphone. Avoid using this mic for broadcast work because it often picks up as much internal noise (tape recorder motor, etc.) as the sound you want. Use a good-quality mic like those mentioned earlier in the chapter on microphones. Some other features that should be part of a professional-quality cassette machine are a VU meter, three heads (rather than two), standard mic inputs, and durable construction. Review Figure 5.11 for additional items that make a good broadcast-quality cassette recorder.

5.15 CASSETTE TAPES

The cassette tape is a small ⅛-inch reel-to-reel tape housed in a plastic case (see Figure 5.13). A short **leader tape** is attached at each end, and both ends of the tape are permanently attached to the reels. When recording onto cassettes, it's important to remember the leader tape because if you're at the very beginning of the cassette, the actual recording will not begin for a few seconds, until you are past the leader tape.

FIGURE 5.11 Broadcast-quality cassette recorder. (Courtesy of TASCAM Division, TEAC America, Inc.)

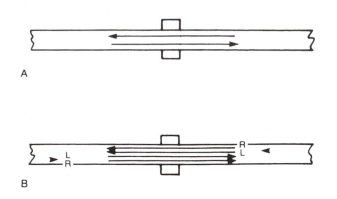

A

B

FIGURE 5.12 Cassette track configurations: a) mono; b) stereo.

Another feature of the cassette is the knock-out tabs on the back edge of the cassette shell. These two little plastic tabs (one for each side of the cassette) allow the recorder to go into the record mode. If you wish to save an important cassette and be sure no one records over the tape, you can knock out the tab. If at a later date you want to record on this cassette, you can put a small piece of cellophane tape over the hole where the tab was, and it will work just as if the tab were there. The audio tape in a cassette is often quite thin so that 30, 60, and 90 minutes of recording time (both sides) can be fit into the plastic shell.

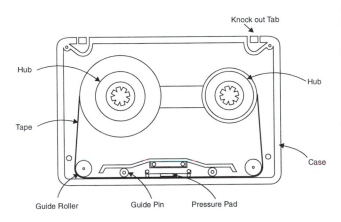

FIGURE 5.13 Internal structure of an audio cassette.

5.16 THE DIGITAL TAPE RECORDERS

Typical reel-to-reel, cassette, and cartridge recorders that have been discussed so far have been the main audio tape recorders used in radio production; however, there are other recorders that are tape-based but add a digital format. For example, a reel-to-reel digital format that is available for two-track and multitrack recorders is the **DASH** (**digital audio stationary head**) system. The two-track systems

use specially formulated ¼-inch audio tape and standard digital technology. Sampling rates are either 48 Khz at standard tape speeds of 7.5 IPS, 15 IPS, and 30 IPS or 44.1 Khz at slightly slower tape speeds (6.9 IPS, 13.8 IPS, and 27.6 IPS). The reduced speed allows the same density of digital data to be put on the tape even at the lower sampling rate. The multitrack systems have found frequent use in the recording studio, but a very high price tag (even the audio tape for the DASH format is very expensive) has kept almost all broadcast stations away from the DASH machines.

There are also digital cartridge recorders, but they use a disc recording medium rather than traditional audio tape, so they will be discussed in a later chapter. Intended as a replacement for the analog cassette, the **DCC** (**digital compact cassette**) is compatible with analog cassettes and will play, but not record, them just like any standard cassette deck. The DCC cassette is the same length and width as the standard cassette but a bit thinner. The hub holes are accessible on the underside of the cassette and then only when a metal shutter (much like the design of a 3½" computer disk) is pushed aside when the cassette is in the recorder. While analog cassettes can be played on DCC recorders, DCC cassettes are not compatible with any other format. All the usual operational controls are present with the addition of some features unique to DCC. For example, text data (such as song titles or liner notes) can be stored on the DCC cassette or subcode information will allow songs to be programmed in any order. Unfortunately, the DCC recorder didn't catch on with either the consumer or the broadcast industry and as of this writing is no longer being manufactured. Although some recorders and tapes may still be available and they will provide digital quality recordings, parts and other supplies will be extremely difficult to find.

5.17 THE DAT RECORDER

The first tape-based digital format to be developed has proven to be the most widely used. The **DAT** (**digital audio tape**) recorder, also known as **R-DAT** (**rotary head digital audio tape**), is technically part VCR, part traditional cassette recorder, and part CD player. As shown in Figure 5.14, although it has the normal tape recorder controls (play, record, rewind, etc.), the tape cassette goes into a slide-out drawer (operated with an open/close control) like many CD players. The DAT unit, like the CD, has several controls for selecting specific songs on the tape. An **AMS** (**Automatic Music Sensor**) button allows the operator to skip forward or backward to the start of the next song. Subcodes can be recorded along with the music on a DAT tape so that the operator can select any individual song by entering in that song's "start ID" or "program number."

A DAT system records with rotating heads, putting ⅛-inch tape in a series of diagonal tracks like the VCR (see Figure 5.15). This is unlike the analog cassette system, which records with a stationary head that puts data on the tape in

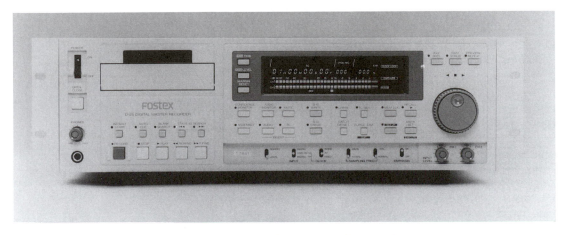

FIGURE 5.14 DAT recorder. (Courtesy of Fostex Corporation of America)

a straight-line (linear) fashion. Tape speed for the DAT recorder is not a factor in the quality of the recording, since only binary data are recorded on the tape; however, with its rotary head drum spinning at 2,000 RPM, the DAT records at an effective tape-to-head speed of over 120 IPS. This high "recording speed" makes it possible to get all the necessary binary data on the tape. The result is the outstanding digital sound quality.

5.18 DAT CASSETTES

The actual DAT cassette tape is similar to a small VCR tape and consists of two little reels encased in a plastic housing about the size of a deck of playing cards (see Figure 5.16). The audio tape is permanently attached to the two reels, and the longest DAT tapes are capable of recording over 2 hours. While cueing time for the CD player is only 1 to 2 seconds, cueing time from one end of the DAT tape to the other is about 40 seconds. This slower access time is of some concern to broadcasters and would probably necessitate

more available DAT sources than CD sources in practical use. On the other hand, cueing a DAT deck from any position to another position ten minutes either side of the current position can be done in ten seconds or less, which compares with the normal run-out time on the audio tape cartridge, making DAT a possible replacement for the cart machine.

5.19 ADVANTAGES OF THE DAT RECORDER

The DAT recorders have the same superior sound quality associated with CDs—exceptional frequency response and signal-to-noise ratio, wider dynamic range, and virtually no wow, flutter, hiss, hum or distortion. In addition, the DAT has been developed with recording capability, and the material recorded can be dubbed almost endlessly without degradation of quality. Even with recordable CDs now available, the DAT has the advantage of a longer recording time—2 hours, or about twice as long as the CD. Perhaps

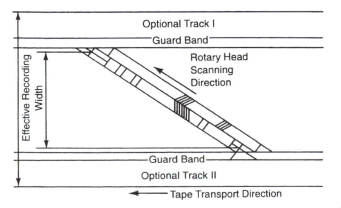

FIGURE 5.15 DAT track configurations.

FIGURE 5.16 Digital audio tape cassette. (Courtesy of TDK Electronics Corporation)

DAT's biggest advantages are its convenience of handling (similar to working with a cassette) and the relative compactness of its portable units. However, DAT's future use is unsure because it faces stiff competition from other digital recording equipment.

5.20 TAPE RECORDER MAINTENANCE

Most tape recorder maintenance is related to the heads themselves. One common problem is, simply, dirty heads. Tape heads can become dirty just from day-to-day use, and, as you'll see in the next chapter, editing audio tape can put unwanted grease-pencil marks on the heads. It's good production practice to clean the tape heads before you begin any production work. The production studio usually has a supply of cotton swabs and head cleaner or denatured alcohol to gently clean any residue off the tape recorder heads. If in doubt about what you're doing, ask for help.

One reason for keeping the tape path clean is to prevent wow and flutter. Both wow and flutter are tape recorder problems that are related to changes in the speed of the tape as it passes through the tape transport. **Wow** refers to slow variations in tape speed, and **flutter** refers to rapid variations. Both problems result in off-pitch sounds that are reflected in their names.

Another maintenance concern regarding tape heads is **alignment**. Basically, alignment is the relationship between the tape as it passes the head and the head itself. There are five adjustments—**azimuth, height, penetration, rotation,** and **zenith**—that can go awry and cause a poor alignment between the tape and head, thus poor frequency response and possible incompatibility with tapes recorded on other machines as shown in Figure 5.17. Alignment problems are definitely best left for the engineer, but it's not a bad idea to be familiar with them.

A final tape recorder maintenance concern is a buildup of magnetism on the tape recorder heads. After extended use, the tape heads will tend to become magnetized permanently. This could result in the heads actually erasing part of the signal that you don't want erased. Usually it will affect the high-frequency signals first. To prevent this, a **demagnetizer** is used to get rid of any built-up magnetism. Usually this is part of the general maintenance done by the engineer, but in some production facilities this type of maintenance, along with cleaning of the heads, is left to the operators. A demagnetizer is brought near to (but not in contact with) the heads, guides, and other metallic parts in the tape path. Since it operates like a bulk eraser, you must be sure to turn it on and off away from the heads (refer to Chapter Eight).

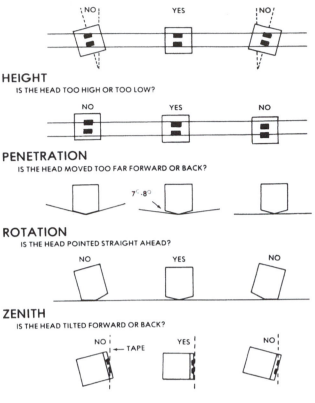

FIGURE 5.17 Head alignment problems. (Courtesy of Lauderdale Electronics)

5.21 CONCLUSIONS

The amount of information contained in this chapter should be an indication to you of the importance of the material covered. You will only become a good production person through practice and use of the equipment described, but if you've absorbed the basic information in this chapter, you have a good solid background for working with the audio tape recorders you'll find in any production facility. You should also have noted that audio tape recorders, like other radio production equipment, are moving from analog to digital. The DAT recorder was introduced as one of the newer and fairly successful digital formats; however, in Chapter Ten you'll learn about several "tapeless" formats that may ultimately replace the traditional audio tape recorders.

Self-Study

■ QUESTIONS

1. Which of the following could be a proper head arrangement for an audio tape recorder, assuming the tape goes from left to right?
   a) erase–record–play
   b) record–erase–play
   c) play–record–erase
   d) erase–play–record

2. If you put a 5-inch reel of tape on a reel-to-reel recorder at $7\frac{1}{2}$ IPS, another 5-inch reel on a recorder at 15 IPS, and a third 5-inch reel on a recorder at $3\frac{3}{4}$ IPS, and start all the recorders at once, which one will run out of tape first?
   a) the reel at $3\frac{3}{4}$ IPS
   b) the reel at $7\frac{1}{2}$ IPS
   c) the reel at 15 IPS
   d) they would all run out of tape at the same time

3. Which of the following prevents crosstalk?
   a) multitrack recording
   b) the pinch roller
   c) leader tape
   d) guard bands

4. A sel sync feature in an audio tape recorder _____.
   a) makes the record head act as a play head
   b) places the erase head after the play head
   c) converts a two-head recorder to a three-head recorder
   d) converts a two-track recorder to a quarter-track recorder

5. If sound is to go from a tape recorder to an audio console, a connection should be made in the tape recorder at _____.
   a) line in
   b) mic in
   c) speaker out
   d) line out

6. Of the tapes in the diagram below, tape B would probably be used on a _____.
   a) cassette recorder
   b) reel-to-reel recorder
   c) cartridge recorder
   d) none of the above

7. The erase head _____.
   a) rearranges the iron particles, so they're in a random pattern
   b) is on during the recording process
   c) closes the gap on the metal core, so the signal jumps to magnetic energy
   d) rearranges the iron particles analogous to the original sound

8. Which of the following represents half-track mono recording?

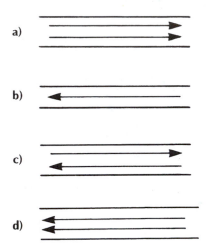

9. Which of the following is an adjustment that changes the amount of amplification given to highs and lows of the sound signal?
   a) digital
   b) bias
   c) equalization
   d) sel sync

10. Which of the following represents quarter-track stereo recording?

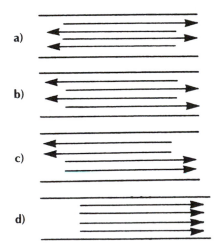

11. Referring to Figure 5.4, which of the following controls the speed of the tape as it passes the head?
   a) feed reel
   b) tape guide
   c) capstan
   d) take-up reel

12. What would happen if a two-track stereo tape were played back on a half- track mono recorder?
   a) both stereo tracks would be heard, but one would be going backward
   b) the tape would be intelligible, but the lower track's material would be missing
   c) four sounds would be heard—two going forward and two going backward
   d) no sound would be heard

13. If the source/tape switch on a tape recorder is in the source position, the VU meter shows _____.
   a) a zero signal
   b) the outgoing signal that has been recorded
   c) the signal that was erased
   d) the incoming signal that is being recorded

14. Which of the following represents half-track mono cassette recording?

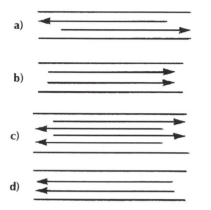

a)

b)

c)

d)

15. What would happen if a stereo cassette recorded on both sides were played on a mono cassette recorder?
   a) two sounds would be heard, one going forward and one going backward
   b) the stereo tape would be intelligible but would not be stereo
   c) there would be a garbled sound
   d) there would be no sound since they're not compatible

16. One difference between a cassette recorder and a cartridge recorder is _____.
   a) a cartridge recorder needs to be threaded
   b) only cassette recorders are used in radio stations
   c) the cartridge tape has no actual beginning or end
   d) only cassette recorders have separate record and play heads

17. Worn cartridge tape will look _____.
   a) shiny
   b) dull
   c) oily
   d) clear

18. Which type of tape must be erased with a degausser?
   a) reel-to-reel
   b) cassette
   c) cartridge
   d) DAT

19. If a tape has a cue tone on it, it will _____.
   a) start automatically
   b) stop automatically
   c) repeat itself
   d) not record

20. The audio tape format used in DAT machines is most similar to the format of _____.
   a) VCR tapes
   b) cassette tapes
   c) reel-to-reel tapes
   d) DCC tapes

21. Which of the following digital formats is compatible with analog cassettes?
   a) DASH
   b) AMS
   c) DAT
   d) DCC

22. Which audio tape recorder is not a closed reel system?
   a) DAT
   b) cassette
   c) cartridge
   d) reel-to-reel

23. If a tape recorder head leans too far to one side, this is a problem related to _____.
   a) rotation
   b) penetration
   c) height
   d) azimuth

24. Which audio tape recorder is for all practical purposes now defunct?
   a) cassette
   b) DCC
   c) DAT
   d) cartridge

25. As a review of audio tape recorders, match the terms in the top list (1, 2, 3 . . .) with their definitions in the bottom list (d, h, r . . .), and then select the correct set of answers from the sequences shown in a, b, or c, or d below.

   1. _____ idler arm
   2. _____ half-track mono
   3. _____ DASH
   4. _____ knock-out tabs
   5. _____ half-track stereo
   6. _____ DAT
   7. _____ capstan
   8. _____ bias
   9. _____ feed reel
   10. _____ ERP
   11. _____ multitrack
   12. _____ crosstalk

   d. a reel-to-reel digital format
   h. a way to remember the head arrangement on tape recorders
   r. a cassette digital format that records diagonally, using a rotating head
   b. a tension control that drops into an off position if the tape breaks
   l. the left, or supply, reel of a reel-to-reel tape recorder
   p. can be used to prevent tape from being recorded on accidentally
   s. controls the speed of tape
   i. an inaudible high-frequency tone used to raise the overall level of a recording
   a. a signal from one track that is picked up on an adjacent track
   o. a track configuration where all the signals go in one direction
   t. a track configuration where one signal is recorded on the top half of the tape and another is recorded on the bottom half, going in the opposite direction
   w. a track configuration where two signals go one way on the top half of the tape, and two go the other way on the bottom half of the tape

   a) 1.b 2.t 3.r 4.p 5.w 6.h 7.s 8.i 9.l 10.d 11.o 12.a
   b) 1.b 2.o 3.d 4.p 5.t 6.r 7.s 8.i 9.l 10.h 11.w 12.a
   c) 1.s 2.t 3.d 4.l 5.w 6.r 7.b 8.a 9.p 10.h 11.o 12.i
   d) 1.b 2.t 3.d 4.p 5.w 6.r 7.s 8.i 9.l 10.h 11.o 12.a

■ ANSWERS

If You Answered A:

1a. Right. And this is the most common head arrangement.

2a. No. (Reread 5.6.)

3a. Wrong. (Reread 5.10 and 5.11.)

4a. Right. This feature allows you to hear previously recorded material that matches with what you are recording.

5a. No. You want an output, not an input. (Reread 5.9.)

6a. Yes. It is $\frac{1}{8}$-inch wide, so it would be used on a cassette.

7a. No. It rearranges them into an orderly pattern. (Reread 5.5.)

8a. No. The signals must go in opposite directions. (Reread 5.10.)

9a. No. This is a recording technique using binary technology. (Reread 5.5.)

10a. Right. You figured the tracks correctly.

11a. Wrong, this is the reel that the tape is placed on, but it doesn't really control tape recorder speed. (Reread 5.8.)

12a. No. At first glance, this looks logical, but it's not. A half-track mono only plays back half a tape at a time. If it played back the full track, it would play back both sounds of its own tapes. All that is played back is the top half of the tape. (Reread 5.10 and 5.11.)

13a. No. There would be a reading on the VU meter. (Reread 5.9.)

14a. Correct. This is the same configuration as $\frac{1}{4}$-inch half-track mono.

15a. No. (Review Figure 5.12; reread 5.14.)

16a. No. Neither needs to be threaded. (Reread 5.12–5.15.)

17a. Right. A worn cartridge tape looks shiny.

18a. No. A reel-to-reel recorder has an erase head that can erase right before new material is recorded. (Reread 5.12.)

19a. No. It must be started by hand. (Reread 5.12.)

20a. Correct. A DAT tape is like a small videotape cassette.

21a. No. This is a reel-to-reel format, so it could not be compatible with cassettes. (Reread 5.16.)

22a. No. A DAT recorder is a closed reel system. (Reread 5.1.)

23a. No. Rotation refers to the head being pointed straight ahead. (Reread 5.20.)

24a. Wrong. Cassette recorders are still utilized in radio production. If you're thinking of a digital cassette recorder, you're on the right track. (Reread 5.14 and 5.16.)

25a. No. You are confused about digital formats. (Reread 5.16 and 5.17.)

If You Answered B:

1b. No. Erase must be before record. (Reread 5.3.)

2b. No. (Reread 5.6.)

3b. Wrong. (Reread 5.8 and 5.11.)

4b. No. You can't move the heads, and you wouldn't want that configuration. (Reread 5.3 and 5.4.)

5b. No. Only microphones should be connected to mic inputs. (Reread 5.9.)

6b. No. (Reread 5.10 and 5.15.)

7b. Right. It is on so that the tape can be erased before it's recorded on again.

8b. No. This is, if anything, backward full-track. (Reread 5.10.)

9b. No. Bias improves frequency response and provides a distortion-free signal. (Reread 5.5.)

10b. No. (Reread 5.10. Pay particular attention to which tracks are used in which direction.)

11b. No. This guides the tape through the recorder mechanism but not at any particular speed. (Reread 5.8.)

12b. Your answer is correct; half-track mono only plays back the top half of the tape.

13b. No. The tape position would do that. (Reread 5.9.)

14b. No. This is like two-track stereo on $\frac{1}{4}$-inch tape. (Reread 5.10 and 5.14.)

15b. Right. Cassette mono tape and stereo tape can each be played on the other recorder and be understood.

16b. No. In fact, cart machines are more common in radio stations than cassettes. (Reread 5.12–5.15.)

17b. No. (Reread 5.13.)

18b. No. A cassette recorder has an erase head that can erase right before new material is recorded. (Reread 5.12.)

19b. Right. Because of the cue tone, it will stop automatically before whatever is next can play.

20b. Wrong. While it's called a cassette tape, it's different from the standard audio cassette. (Reread 5.15 and 5.18.)

21b. No. You are way off. Automatic Music Sensor is not even a tape format. (Reread 5.16 and 5.17.)

22b. No. A cassette recorder is a closed reel system. (Reread 5.1.)

23b. No. Penetration refers to the head being too far forward or backward. (Reread 5.20.)

24b. Right. The DCC or digital cassette recorder technology never really caught on with consumers or broadcasters.

25b. No. You are confused about track recording configurations. (Reread 5.10, 5.12, and 5.14.)

If You Answered C:

1c. No. Erase must be before record. (Reread 5.3.)

2c. That's right. It goes through the recorder the fastest and so would finish first.

3c. Wrong. (Reread 5.11 and 5.15.)

4c. Wrong. Adding a head is a much more complex process. (Reread 5.3 and 5.4.)

5c. No. This could go to an external speaker, but not an audio console. (Reread 5.9.)

6c. No. (Reread 5.10, 5.13, and 5.15.)

7c. Wrong. This answer really doesn't make sense. You are probably confused about many of the terms in this chapter. (Reread the entire chapter.)

8c. Good. Your answer is correct.

9c. You are correct.

10c. No. (Reread 5.10. Pay particular attention to which tracks are used in which direction.)

11c. Right. This, in conjunction with the pinch roller, pulls the tape through at a uniform speed.

12c. No. There are only two tracks on two-track stereo, so it couldn't play back four. (Review Figure 5.11; reread 5.10 and 5.11.)

13c. No. Nothing shows the signal erased. (Reread 5.9 very carefully.)

14c. No. You are way off; this has four tracks. (Reread 5.10 and 5.14.)

15c. No. (Review Diagram 5.12; reread 5.14.)

16c. Right. The cartridge tape is a continuous loop.

17c. No. The tape has a lubricant, but it doesn't look oily. (Reread 5.13.)

18c. Right. A cartridge machine has no erase head.

19c. No. The purpose of the cue tone is to prevent the tape from repeating itself. (Reread 5.12.)

20c. Wrong. Although there are tiny reels inside the DAT cassette, it is far from being like an open reel audio tape. (Reread 5.8 and 5.18.)

21c. Close, but not right. You chose the wrong cassette format. (Reread 5.16 and 5.17.)

22c. No. A cartridge recorder is a closed reel system. (Reread 5.1.)

23c. No. Height refers to the head being too high or too low. (Reread 5.20.)

24c. Wrong. The DAT or digital audio tape recorder is used in radio production, remote recording, and the recording studio. (Reread 5.16 and 5.17.)

25c. You are confused about many of the tape recorder parts and functions. (Reread the entire chapter.)

If You Answered D:

1d. No. This is not a common head arrangement. (Reread 5.3.)

2d. You might be confused because they all have the same size reel, but you're not correct. (Reread 5.6.)

3d. Right. Guard bands prevent crosstalk.

4d. Wrong. Some recorders do have a switch to accomplish this, but it has nothing to do with sel sync. (Reread 5.4 and 5.10.)

5d. Right. Line out would go to the audio console.

6d. Wrong. One of the other answers is correct. (Reread 5.10, 5.13, and 5.15.)

7d. No. You're getting this confused with the record head. (Reread 5.5.)

8d. No. You're not correct. (Reread 5.10.)

9d. Wrong. This allows you to record one track in synchronization with a previously recorded track. (Reread 5.4 and 5.5.)

10d. No. This is multitrack recording, in this case four-track. (Reread 5.10.)

11d. No. As the tape goes past the heads it is taken up by this reel, but it doesn't really control the speed. (Reread 5.8.)

12d. Wrong. While there is not full compatibility between two-track stereo and half-track mono, some sound would be heard. (Reread 5.10 and 5.11.)

13d. Correct. "Source" shows the source.

14d. No. This is not the correct track configuration. (Reread 5.10 and 5.14.)

15d. No. Stereo and mono cassettes are compatible. (Reread 5.14.)

16d. Wrong. Both cartridge and cassette recorders can have separate play and record heads. (Reread 5.12–5.15.)

17d. No. If a cart tape was clear, it wouldn't have any oxide particles on it at all—it would be beyond worn! (Reread 5.13.)

18d. Wrong. DAT tape is erased during the recording process. (Reread 5.12.)

19d. No. A cue tone would not prevent a tape from being recorded. (Reread 5.12.)

20d. No. While there are some similarities here, there is a better answer. (Reread 5.16 and 5.18.)

21d. Correct. DCC can play back analog cassette tapes.

22d. Yes. A reel-to-reel recorder is an open-reel system.

23d. Right. Azimuth refers to how the head leans.

24d. Wrong. The audio cartridge recorder may not have the importance it once did, but it's still used in radio production. (Reread 5.12 and 5.16.)

25d. Correct. You have now completed the section on audio tape recorders.

Projects

■ PROJECT 1

Dub taped material between cassette and reel-to-reel recorders.

Purpose

To make sure you are able to do even, clear dubbing, which is one of the most common exercises done in production work.

Advice, Cautions, and Background

1. The material you are to dub (which can be taken from the CD that accompanies this text) was *not* recorded properly in that it is not all at the same level. You are to make it as much at the same level as possible. On professional equipment, there are sophisticated meters to help you. On less expensive equipment, you will have to do this by practicing a few times and getting the feel of it.
2. If you're unsure of what you're doing, get help. Don't ruin the equipment.

How to Do the Project

1. Make sure you know how to operate your cassette recorder. If in doubt, ask the instructor.
2. In the same way, make sure you know how to operate the reel-to-reel recorder.
3. Transfer the material from the CD to your reel-to-reel recorder without making any level adjustments. If you have difficulty with this, ask your instructor for help.
4. Listen to the material, noting where the level changes are, and decide on a strategy for dubbing it so that it's all at one level.
5. Rewind the tape, and connect the cable from the proper audio output on the reel-to-reel to the proper audio input on the cassette recorder. If your equipment is already connected through an audio console, just be sure the switches and knobs are set properly so that you can record from reel-to-reel to cassette.
6. Place a tape in the cassette recorder.
7. Put the reel-to-reel recorder in play. Monitor the recording, and, at the same time, adjust the volume on one of the recorders so the level will be even.
8. Rewind the reel-to-reel recorder, and set the cassette recorder so that it will record and the reel-to-reel recorder so that it will play. The sound will then go from the reel-to-reel to the cassette recorder.
9. Practice your dub several times, adjusting levels until you get the feel of how much you need to vary the volume. Then make the dub.
10. Now you're ready to make your dub from cassette back to the reel-to-reel recorder. Place a different tape on the reel-to-reel recorder.
11. If necessary, change cables so that one goes from the proper audio output on the cassette to the proper audio input on the reel-to-reel. (Again, you may just have to set the proper switches and knobs on your audio console.)
12. Decide on a volume for both recorders, and put the cassette recorder in play and the reel-to-reel recorder in record.
13. Make your dub. This time you shouldn't have to adjust levels. Listen to the dub to make sure it recorded properly.
14. Label the tape with your name and Tape Dub. Turn it in to the instructor to get credit for this project.

■ PROJECT 2

Record several generations of an audio signal, using analog and digital recording processes.

Purpose

To enable you to hear the loss of signal quality inherent in the analog recording process, but not in digital.

Advice, Cautions, and Background

1. This project assumes that you have a DAT or other digital recorder in your production facility. If you don't, you can still do the exercise, using only analog recorders. Compare the dubbed sound with the original CD sound.
2. As you make the analog recordings, watch the levels and other elements that might introduce a poor quality recording due to operator error.

How to Do the Project

1. Review the section on digital and analog recording in Chapter Four.
2. Use a CD as your original sound source.
3. Make a dub of the CD onto an analog audio tape recorder. This will be the first-generation recording.
4. Now, using *the tape that you just made* as the sound source, make another analog tape recording. This will be the second-generation recording.
5. Now, using *the second-generation recording* as the sound source, make another dub, or the third-generation recording.
6. Listen to and compare the three recordings you just made with the original sound source (the CD). You should notice a loss of sound quality plus increased noise and distortion, among other things.
7. Repeat steps 2 through 6, using a digital audio tape recorder.
8. Listen to and compare the digital recordings with the original sound source.
9. Write a one-page report that summarizes what you feel are the differences between the analog and digital recordings.
10. Turn in your report labeled ANALOG/DIGITAL RECORDING and both the analog and digital recordings to your instructor for credit for completing this project.

■ PROJECT 3

Label diagrams of the various types of reel-to-reel, cassette, cartridge, DAT, and/or other tape recorders in your production facility.

Purpose

To familiarize you with the various functions of your production room tape recorders.

Advice, Cautions, and Background

1. No two brands of recorders are exactly alike, so you may find buttons or switches not specifically discussed in this chapter. Ask your instructor for help if you can't decide how to label something.
2. If there are repetitive switches and knobs, you don't need to label all of them, but make sure that you understand and have indicated all the main functions.
3. If you have many different brands of one type of recorder (e.g., Sony, Denon, Tascam cassette recorders), you don't need to draw all of them. One or two will suffice.
4. Use a separate piece of paper for each recorder so that you have plenty of room to write.
5. You'll be judged on the completeness and accuracy of your drawings. You won't be graded on artistic ability, but be as clear as possible.

How to Do the Project

1. Sketch the audio tape recorders found in your production room. If necessary, draw several sides of the recorders so that you have included all the important function switches.
2. Label the basic parts: heads, speed controls, counters, reels, tape guides, capstan, pinch roller, idler arm, play, record, fast forward, pause, VU meters, pots, source/tape switches, etc.
3. Make sure you know what each of these does.
4. If possible, give the brand name and model number of each tape recorder.
5. Put your completed drawings together with a page titled AUDIO TAPE RECORDERS. Remember to put your name on each sketch.
6. Give this packet to your instructor for credit for this project.

CHAPTER SIX

Audio Tape Editing

Information

6.1 INTRODUCTION

As you'll see in Chapter Ten, the techniques for audio tape **editing** are changing, but for now, as a radio production person, one of the more important skills you need to learn is "old-fashioned" audio tape editing. From creating a music bed to adding sound effects to editing out vocal "fluffs," or mistakes, tape editing is a day-to-day part of radio production work. And lest you think you're learning outdated skills, many of the basic procedures of audio tape editing are the same, whether you're using "cut and splice" methods or manipulating the latest digital editing system. Before looking at the techniques of audio tape editing, let's examine the material you'll be working with—the audio tape itself.

6.2 PHYSICAL MAKEUP OF AUDIO TAPE

Audio tape for a radio production person is like paper to a writer. In other words, it's the storage medium on which various ideas are put down and then manipulated. Both the writer and the radio person can erase unwanted segments or edit long segments into shorter, workable concepts, but for the radio person it all starts with getting something onto tape first.

Although we see audio tape as merely a thin ribbon of tape, its physical makeup actually consists of three layers: a **plastic base** sandwiched between a **backing layer** and a **magnetic layer** (see Figure 6.1). The back coating provides traction as the audio tape moves through a tape recorder transport, and it also provides protection from tape breakage and print-through.

Print-through is defined as the transfer of the magnetic signal on one layer of tape to the magnetic signal on the next layer of tape either above it or below it on the reel. Think of the sandwich concept again, and visualize a jelly sandwich stacked on top of a peanut butter sandwich. If the jelly soaks through the bottom piece of bread and onto the top piece of bread on the peanut butter sandwich, print-through has occurred. It's most audible when one of the tape layers contains a very loud sound, and the adjacent layer contains a soft sound. While most modern audio tape

is not very susceptible to print-through, you can prevent it by recording at normal levels (i.e., avoid recording "in the red") and by avoiding using thin (less than 1.5 mil) tapes. Print-through can occur when audio tape is stored on a reel for long periods of time without being played. Winding and rewinding stored tape occasionally may also help prevent print-through.

Are digital tapes, such as DAT, subject to print-through? Yes, they are, but the digital playback system simply ignores it. As noted above, on analog tapes, the print-through signal may be clearly audible when it's loud, and the original audio signal is soft. However, in digital recording systems, the playback electronics detect only the presence of the digital signal at any given time and when the signal is too low (such as print-through would be), then the playback system doesn't detect it at all.

The plastic base (the middle of our sandwich) in modern audio tape is polyester. This produces a strong tape that rarely breaks; however, it will stretch if pulled too hard. Early audio tape used an acetate base that broke cleanly, but it was susceptible to temperature and humidity effects. Often, polyester audio tape is **tensilized,** or prestretched, to prevent stretching problems during day-to-day use.

The top of the audio tape sandwich is an iron-oxide coating. This coating is composed of tiny slivers of magnetic oxides that are capable of storing an electromagnetic signal that's analogous to the original sound made during the recording process. Instead of an iron-oxide coating, modern audio tape offers different magnetic coatings, such as chromium dioxide, for better recording characteristics.

6.3 PHYSICAL DIMENSIONS OF AUDIO TAPE

Audio tape is only about one-thousandth of an inch thick. Audio tape thickness is measured in mils, or thousandths of an inch. Most tape used in radio production work is either 1 mil or 1.5 mil; the latter is preferred by broadcasters. There are thinner audio tapes available, such as those used in audio cassettes. The advantage of thinner tape is that you can get more tape on a reel, which gives you more playing time. But thin tape also stretches more easily, is more susceptible to print-through, and is very difficult to handle in

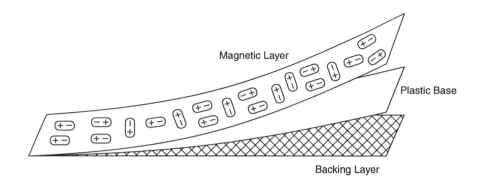

Magnetic Layer

Plastic Base

Backing Layer

FIGURE 6.1 Audio tape "layers."

splicing. Tape specifically designed for digital recording is usually thinner than standard audio tape because it only has to store digital data on it.

The other physical dimension of audio tape that concerns production people is its width. The standard width for reel-to-reel audio tape is ¼ inch. Cassette tape is usually considered to be ⅛-inch wide (actually .15 inch), and some multitrack recorders found in the radio production studio use ½-inch tape. Most wider audio tape (1 and 2 inch) is only found in the recording studio.

6.4 CASSETTE TAPE TYPES

Audio cassettes are classified according to the kind of tape formulation used in the cassette. The magnetic material used in standard cassettes is ferric oxide ("rust" in its natural form), and these cassettes are designated as **Type I** tapes. **Type II** tapes use a chromium-dioxide or a chrome-equivalent formulation. A **Type III** cassette tape, which was a dual-layer ferri-chrome tape, existed briefly but is no longer available. Cassette tapes that use a pure metal magnetic material rather than an oxide compound are known as metal or **Type IV** tapes.

Each of the cassette tapes requires different amounts of **bias** (during recording) and **equalization** (during playback). Type I tapes are called "normal bias" tapes because they use a bias current near the usual level for reel-to-reel tapes. Type II tapes are "high bias" because they use a bias level about 50 percent greater than Type I tapes, and Type IV tapes use an even higher bias. Cassette recorders usually employ one of two kinds of playback equalization: 120-microsecond ("normal" equalization) for Type I tapes and 70-microsecond for Type II and Type IV tapes. Some cassette decks automatically set the bias and equalization, using special sensing cutouts on the plastic cassette shell, but many require the operator to set the correct tape type.

Type I and Type II tapes are most frequently found in the radio production studio situation. While there isn't a huge quality difference between them, Type II tapes should produce a slightly better recording. Metal tapes (Type IV) actu-ally provide the best quality but are expensive for day-to-day broadcast use.

6.5 AUDIO TAPE DEFECTS

Although audio tape has proven to be an excellent working medium for radio production, it's not perfect. Perhaps the biggest problem with audio tape is signal loss due to drop-out. **Drop-out** is a defect in the oxide coating that prevents the signal at that point from being recorded at the same level, or at all. Drop-out is a problem that occurs in the manufacturing process of audio tape, but it can also be caused by flaking of the oxide coating due to heavy use or abuse of the tape.

Other tape defects or problems that you're likely to encounter during production work include adhesion, scattered winds, curling, and cupping. **Adhesion** occurs when tape has been stored on reels for some time, and one layer sticks to another as it's unreeled. Tape stored in humid conditions is more likely to experience adhesion problems. Occasionally tape will wind unevenly onto a reel, leaving some edges of the tape exposed above the rest of the tape on the reel. This is known as a **scattered wind**, and you must be careful not to damage the exposed tape strands. Both **curling** (tape twisting front to back) and **cupping** (tape edges turning up) are problems due to a poor binding between the plastic base and the oxide coating. See Figure 6.2 for examples of these common tape defects. Almost every tape problem can be avoided by using high-quality tape for all your radio production work.

6.6 REASONS FOR EDITING

Why edit audio tape? The answers are relatively obvious. For example, rarely will you produce the vocal track for a commercial exactly the way you want it on the first try. Editing gives you the ability to eliminate mistakes. You can edit out fluffs and keep only the exact words you want. Other production work may require you to edit out exces-

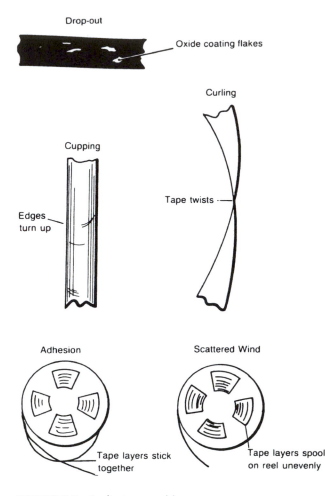

FIGURE 6.2 Audio tape problems.

sive pauses or "uhs" from an interview or language not allowed by the FCC from a piece of news tape.

In addition to eliminating mistakes, editing allows you to decrease the length of production work. Radio requires exact time for commercials, news stories, etc., and editing can keep your work to the exact lengths required. Audio tape editing also allows you to record out of sequence. You might be putting together a commercial that uses the testimonial of several customers, and the one that you want to use first in your commercial may not have been recorded first. Editing allows you to easily rearrange the order or use just a portion of what you actually taped.

6.7 TYPES OF EDITING

There are traditionally two terms associated with audio tape editing: splicing and dubbing. **Splicing** normally refers to physically cutting the audio tape, taking a portion out and splicing the remaining pieces back together. It's almost always done with a reel-to-reel recorder because the

tape in a cart or cassette is essentially inaccessible. **Dubbing** is often considered electronic editing, by which portions of one tape are copied onto another tape. Although splicing and dubbing have been the standard ways to edit audio tape, digital technology promises to make digital audio editing as common in the future as electronic video tape editing is today. **Digital audio editing** refers to any system that uses hard-disk storage and computer software to manipulate audio with either a PC computer or proprietary editing equipment.

6.8 TOOLS OF THE TRADE

Before beginning to learn the technique of audio tape editing, let's assemble the tools and supplies that we are going to need. The basic tools of the trade for splicing are a grease pencil, a splicing block, a razor blade, and splicing tape.

This type of editing requires physically marking on the tape the points where you're going to cut it. A white or yellow **grease pencil**, or **china marker**, is used to accomplish this. Some tape recorders found in the production studio have a built-in marker, and in a pinch you can carefully use a soft lead pencil, but most often you'll use a grease pencil. The side of the audio tape that is marked is the back, or unrecorded, side. When marking, make sure none of the grease pencil gets on the heads or on the front (recorded) side of the tape. The crayonlike substance in a grease pencil can clog the head, so it should not come in contact with it.

A **splicing block** is a small metal or plastic block (see Figure 6.3) with a channel to hold the audio tape and two grooves to guide the razor blade when cutting the tape. The channel has tiny lips at its edges so that the audio tape snaps down into it. The tape can easily be slid in the channel by the operator, but it won't move on its own. The cutting grooves are at 45 degrees and 90 degrees to the audio tape. For almost all production work, you'll use the diagonal cut. Although broadcast supply companies offer industrial-grade razor blades, almost any standard *single-edged* razor blade will work for cutting audio tape. Be careful! Razor blades are sharp and will cut your fingers as easily as they cut audio tape.

Splicing tape is commercially available and is specially designed so that its adhesive material does not soak through the audio tape and gum up the heads of the tape recorder. Never use cellophane tape or other office tape to do editing work. Splicing tape is slightly narrower than audio tape ($7/32$ inch versus $1/4$ inch) so that any excess adhe-

FIGURE 6.3 Splicing block. (Courtesy of Xedit Corporation)

sive material will not protrude beyond the edge of the audio tape. In addition to rolls of splicing tape, you can get splicing tabs—precut pieces of splicing tape on applicator strips, designed to make the splicing process easier.

There are audio tape splicers that combine a splicing block, razor blade, and splicing tape dispenser into one unit. Not only do they allow you to perform normal splicing techniques, but they also allow you to trim the edges of the spliced portion of the audio tape slightly. While this ensures smooth, clean edits, it's also possible to lose the edge tracks on the tape or cause misalignment when the tape passes the heads. For whatever reason, these splicers have not found universal acceptance in the radio production studio, so you may or may not run across them in your production work.

6.9 MARKING EDIT POINTS

Audio tape editing is really a two-step process: marking the edit points and then splicing. Since audio tape passes through the recorder from left to right, sounds are recorded on the tape in the same manner. For example, the phrase "editing is really a two-step process" would be recorded on audio tape in this manner: "ssecorp pets-owt a yllaer si gnitide." The letters would appear on the tape just as they do on this page, looking at them left to right, but the rightmost word ("editing") would be recorded first. If we wanted to edit out the word "really" in this phrase, we would make two edit points, one on each side of the word. In tape editing, the edit point is always made in front of the word (or sound) that you wish to edit out on the one side and in front of the word (or sound) that you're leaving in on the other side. Looking at our example again, you would mark just in front of the r in "really" and just in front of the "a." It's important that you always mark in front of words to maintain the proper phrasing of the speech. If you had marked just before the r in "really" and just after the y to edit out the word, you would have lengthened the pause between "is" and "a" on your edited tape. Often this will be unnatural sounding and noticeable to the listener.

You already know (from the previous chapter) that the heads in an audio tape recorder and their functions are signified by "ERP" or "erase-record-play." For marking purposes, you are only concerned with the playback head because this is where you'll make your mark. Look at the heads on the recorder that you'll be doing your editing on (see Figure 6.4). Find the playback head, and then find the dead center of that head. Remember that audio tape will be covering the actual head gap, so look for some kind of reference above and below this portion of the head that will allow you to find this spot consistently. Now take "AIM." Every time you make an edit mark, make it slightly *to the right* of dead center. (Marking to the right compensates for the split second it takes you to actually press the stop button when listening to the audio tape to find where to make an edit.) What you are doing is establishing an "*actual indicator mark*"! Your AIM will be exactly the same time after time, and that means your edits will be the same time after time. Your goal should be to make the same edit over and over and have each one sound the same. That is uniformity in editing, and that's a skill you want. You may want to read over this section again. Make sure you understand it because if you mismark an edit, even if your splice is perfect, it won't sound right.

6.10 STEPS IN SPLICING TECHNIQUE

If you've marked the audio tape at two points, it's time to perform the actual splice. Normal splicing technique follows these steps:

A. *Position the tape at the first edit mark in the splicing block.* (See Figure 6.5A.) Remember, the unrecorded side of the tape should be facing up in the splicing block, and the edit mark should be exactly at the 45-degree cutting groove. The diagonal groove is used for splicing for two reasons. First, it provides more surface for contact with the splicing tape at the point of the edit, and thus a stronger bond or splice is made. Second, and more importantly, the diagonal cut provides a smoother sound transition. For example, if you are splicing together two pieces of music, rather than an abrupt change from one piece to another at the edit (such as a 90-degree splice would produce), you have a short blend of the

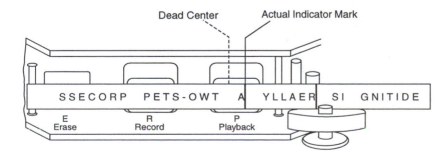

FIGURE 6.4 Tape recorder head assembly.

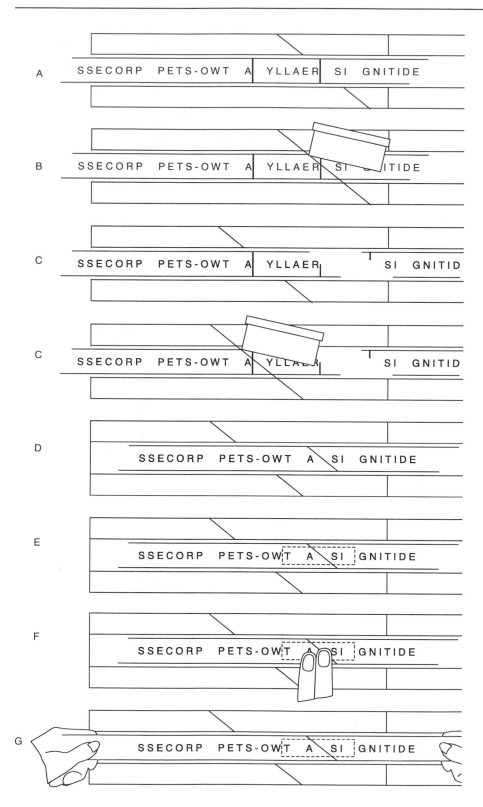

FIGURE 6.5 (a) Position tape at first edit mark. (b) Cut tape at first mark. (c) Position and cut tape at second edit mark. (d) Butt tape ends together. (e) Apply splicing tape. (f) Smooth out splicing tape. (g) Remove tape from splicing block.

music pieces at the edit. Just cut two pieces of tape—one at 45 degrees and one at 90 degrees—set them side by side, and you'll see why the diagonal cut works best for splicing.

B. *Cut the tape at your first mark.* (See Figure 6.5B.) It isn't necessary to saw away with the razor blade or apply excessive downward pressure. A simple slicing motion with the blade through the groove should cut the tape cleanly. Pressing down too hard on the blade will merely make the blade dull faster. If the tape doesn't seem to want to cut, the razor blade is already dull, and you should get a new one. You can tell that a razor blade is getting dull when it doesn't slice cleanly through the audio tape. A ragged cut makes it difficult to align the tape for a good splice. Frequently changing razor blades will also prevent them from becoming magnetized. Because the metal blade cuts through the magnetic layer of the audio tape, over a period of time it will become magnetized, and continued use may put a click on the audio tape or degrade the recorded signal at the point of the edit. Some production people demagnetize the razor blade on the bulk eraser prior to production work, but it's not necessary if you change blades on a regular basis.

C. *Repeat steps A and B at the second edit mark.* (See Figure 6.5C.) Remove the unwanted piece of audio tape, but don't discard it yet. It's good production practice to hang on to cut-out tape until after you're sure the splice has been accomplished as you want it. If you've erred in your AIM, it's possible (although difficult) to splice the cut-out piece of tape back in and try the splice again.

D. *Butt the remaining tape ends together.* (See Figure 6.5D.) Move both pieces of the tape slightly left or right so that you don't butt them together directly over the cutting groove.

E. *Apply the splicing tape on the edit.* (See Figure 6.5E.) If you're using splicing tape from a roll, a piece about ¾ inch in length is ample. Splicing tabs are precut at the appropriate length. The splicing tape should be centered at the edit. Make sure it's positioned straight along the channel of the splicing block. Remember, the splicing tape is narrower than the audio tape, so it should not protrude over either edge of the audio tape.

F. *Smooth out the splicing tape.* (See Figure 6.5F.) Be sure to get air bubbles out from under the splicing tape for a strong bond. Rubbing your fingernail over the splice will usually take care of this.

G. *Remove the audio tape from the splicing block.* (See Figure 6.5G.) Never remove the audio tape from the splicing block by grasping one end and lifting. The lips on the edges of the channel will damage the audio tape. The proper procedure for removing audio tape from the splicing block is to grasp both ends of the tape just beyond the splicing block, apply slight pressure to the tape by pulling your hands in opposite directions, and lift straight up. The tape will pop out of the block, and you will have completed your splice.

Thread the tape on your recorder, and listen to the edited tape. If it came out as you wanted, you can discard the unwanted tape section. Sometimes you may find it necessary to shave a piece of the edit by splicing off one edge or another of the tape. If you've made a good edit mark, you'll rarely have to do this.

6.11 SPLICE DEFECTS

Beginning audio tape editors often encounter problems with their first few splices. These are usually overcome with practice and experience, but it's not uncommon to see splicing errors in manipulating the splicing tape itself and in manipulating the audio tape. (See Figure 6.6 for examples of common splicing problems.) One of the most common problems with splicing tape manipulation is using too much; a piece of splicing tape that's too long is difficult to position properly on the audio tape and makes the tape too stiff at the edit, which prevents proper contact with the tape recorder heads. On the other hand, a piece of splicing tape that's too short may not hold the audio tape together during normal use. Another cause of an edit not holding is the splicing tape being poorly secured on the audio tape. For example, air bubbles or dirt or grease (from fingerprints or excessive marking with the grease pencil) may be under the splicing tape, preventing it from adhering to the audio tape. Other problems arise when the splicing tape is put on crooked; a portion of the splicing tape hangs over the edges of the audio tape, making it impossible for the tape to glide through the tape recorder transport properly.

One of the most common problems with audio tape manipulation during the editing process is leaving a gap as you butt the two tape ends together. Obviously, a gap at the edit point will be heard as an interruption of sound or too long a pause. On the other hand, if you overlap the two tape ends as you butt them together, you'll get the same effect as if you had mismarked the edit point; in other words, the splice won't occur where you thought it would. A less common problem is lining up the pieces of audio tape incorrectly so that one piece of tape is higher or lower than the other as you butt them. If you're using a splicing block, the channel prevents this from happening; however, some editing is done without the benefit of the splicing block, and one can misalign the two pieces of tape. This prevents the tape from gliding through the tape transport properly.

6.12 LEADER TAPE

Leader tape is often used in the editing process. Leader tape is plastic or paper tape that does *not* have a magnetic layer. It is sometimes clear plastic, but more often it's colored as well as marked, or timed, in 7½-inch segments. This timing enables the operator to use leader tape to accurately cue up an audio tape. Leader tape is usually put at the beginning of an audio tape to mark the exact start of the recorded sound. (In contrast with audio tape editing, leader tape is normally spliced onto audio tape using the 90-degree cutting groove.)

WITH THE AUDIO TAPE

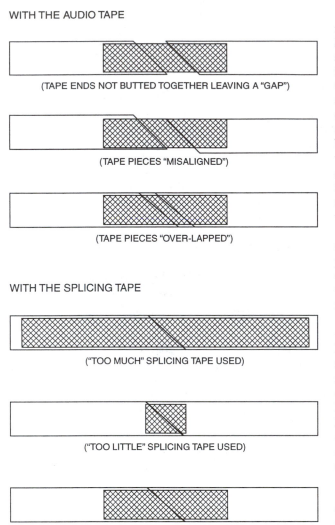

(TAPE ENDS NOT BUTTED TOGETHER LEAVING A "GAP")

(TAPE PIECES "MISALIGNED")

(TAPE PIECES "OVER-LAPPED")

WITH THE SPLICING TAPE

("TOO MUCH" SPLICING TAPE USED)

("TOO LITTLE" SPLICING TAPE USED)

(A GOOD SPLICE)

("AIR BUBBLES" OR "DIRT" TRAPPED UNDER THE SPLICING TAPE

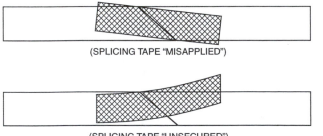

(SPLICING TAPE "MISAPPLIED")

(SPLICING TAPE "UNSECURED")

FIGURE 6.6 Splicing problems.

The beginning of a tape is called the **head** (tape is normally wound on a reel "**heads out**"), and white or green leader tape is commonly used to signify this. Another reason for putting leader tape at the beginning is to use it to thread the tape recorder and avoid wear and tear on the actual recorded portion of the audio tape. Leader tape can also be written on; titles or notes can be put at the beginning of a tape this way. The end of an audio tape is called the **tail,** and if leader tape is put on the end of a tape, it will usually be yellow or red (often the leader tape is actually white, with the timing marks being the appropriate color). Some audio tape (especially that used for prerecorded programs) is stored on reels "**tails out,**" or tails at the end of the recorded material. To play back a tape that is tails out, you would have to rewind the tape first.

In addition to being used at the beginning and end of an audio tape, leader tape is often used to separate various program segments. It's easier to cue up a tape when the operator can see the segments as they pass through the tape transport (even in fast forward) rather than having to listen for pauses between the segments.

As mentioned in the previous chapter, cassette tapes also usually have leader tape at both ends of the tape reels. This keeps the actual audio tape within the plastic shell and not exposed at the open "head area" of the cassette. In addition, if the hub of the cassette (where the tape is attached) is slightly "out-of-round," it can cause sound quality problems at the beginning and end of the tape; however, the leader tape provides a cushion effect that helps prevent any adverse effect. Cassette leader tape is also used to activate an auto-stop or auto-reverse mechanism on some tape recorders.

6.13 DUBBING

As mentioned in 6.7, dubbing is another form of audio tape editing. This electronic editing requires the use of two tape recorders. As you dub or copy from one to the other, the "master" tape recorder is in the play mode, and the "slave" tape recorder is in the record mode. Using our previous example ("editing is really a two-step process"), you would dub "editing is" from the master to the slave tape recorder, and then stop both. Next, you'd cue up the master tape recorder past the word "really" and then dub "a two-step process" from the master to the slave tape recorder. Electronic editing requires a good deal of coordination because you manipulate the two machines so that they start at the same time. Usually electronic editing produces a glitch at the edit point that can range from barely noticeable to terrible, depending on the tape recorders used. As electronic editing becomes more common with digital recording, it will be computer controlled like videotape editing; then audio tape dubbing can be precise.

Straight dubbing is frequently used in radio production work for reasons other than editing. Often a master tape is dubbed onto a working tape before splicing so that the original tape isn't cut during the editing process. You can dub from one tape format to another. For example, you may have a news tape that was recorded on cassette, but you want to put it on reel-to-reel so that you can edit it down or perhaps equalize it to improve quality. Finally, dubbing can be used simply to make duplicate copies of any existing tape.

6.14 EDITING WITH AUDIO CARTS

Editing—or, more correctly, dubbing—can be done with audio tape cartridge recorders if the recorder has **cue defeat** capability. Many cart recorders have a switch that when set for cue defeat will not put a cue tone on the tape at the beginning of the recording. This would allow you to assemble several segments back to back on the cart by recording the first segment (with cue), stopping the cart, setting the cart in record mode again (with cue defeat), recording the next segment, and so on. When played back, the recording would sound like one continuous recording and, of course, would stop at the beginning cue tone after playing.

6.15 DIGITAL AUDIO EDITING

With the advent of digital audio workstations and software that allows you to manipulate audio on a PC computer system, another form of editing—digital audio editing—has been added to the splicing and dubbing techniques. There are many different systems currently available that allow digital editing; however, there is no one standard technique for doing so. To gain an understanding of digital audio editing, we'll look at one system that uses a standard PC computer, digital sound card, and a software program—

Software Audio Workshop Plus—much like the system shown in Figure 6.7. Even though some procedures and some terminology will be different, many of the basic principles would apply to any equipment that's capable of digital audio editing. Much like dealing with the audio console, an understanding of one piece of equipment will give you a good idea for the operation of most of the other available equipment.

The first step in any type of editing is to have the audio you want to edit on the tape, or hard disk in this case. Our system has a RECORD screen that allows us to put audio into the system in a digital format. Controls on this screen (see Figure 6.8) look much like any recorder with play, record, and stop buttons, VU meters, volume controls, and timing windows. Of course, you manipulate the "buttons" and everything else with a few mouse clicks. In this type of system, audio is usually fed into the computer through the audio console, but it can also be put directly into the computer. Audio is recorded in the digital audio editor as a soundfile. To parallel our earlier example, let's record "editing is really a two-step process" into the system and label this soundfile as EDIT.WAV.

Once our audio is recorded, we can pop up a new screen (see Figure 6.9) that allows us to actually edit the soundfile. This SOUNDFILE screen shows a waveform, or a visual representation, of the sound of the recorded audio. Editing is as easy as putting an electronic "mark" at the beginning and ending of the audio section you want to edit out. Looking at our example, one mark would be put just before the *r* in "really" and another mark would be put just before the word "a." The area between the two marks is called a **region** and will become shaded on the screen as shown in Figure 6.9. You can play just the marked region to be sure you have placed the edit marks exactly where you want them.

FIGURE 6.7 PC-based digital audio editing system. (Courtesy of Micro Technology Unlimited)

FIGURE 6.8 Record screen. (Courtesy of Innovative Quality Software)

You can also "zoom in" on the waveform picture to move the edit marks to very precise points on the audio. The mouse controls movement of the edit points easily. Once you are sure you have the two edit marks where you want them, you can just "cut" out the shaded section, or, in other words, complete the edit with a mouse click or two. You can usually preview the edit to make sure it sounds exactly as you want, and, of course, digital editing is nondestructive and allows you to "undo" previous steps so that you can correct mismarks or try other versions of the edit.

We've only touched the surface of what a digital audio editor can do, but it should be obvious, when comparing this procedure with the steps involved in traditional splicing, that digital audio editing is the more accurate, quicker,

and easier technique. In later chapters, we'll look at other ways to manipulate audio with a digital audio editor.

6.16 CONCLUSIONS

This chapter is an important one in your development as a radio production person. You might want to review it again before attempting the Self-Study or Projects sections. We have concentrated mainly on the techniques of audio tape editing, and not so much on the aesthetics. There are some important considerations in the aesthetics area, and they can be developed after you've had time to master the basics.

FIGURE 6.9 Editing screen. (Courtesy of Innovative Quality Software)

Self-Study

■ QUESTIONS

1. Which layer of audio tape provides traction as the tape moves through the tape recorder transport?
   a) backing layer
   b) plastic base
   c) magnetic layer
   d) oxide layer

2. Which of the following is the term for an unwanted effect of the magnetic signal involving adjacent layers of audio tape?
   a) tensilize
   b) drop-out
   c) adhesion
   d) print-through

3. Which type of audio tape is preferred by broadcasters?
   a) polyester
   b) acetate
   c) paper
   d) Type IV

4. Which thickness of audio tape is preferred by broadcasters?
   a) .5 mil
   b) 1.0 mil
   c) 1.5 mil
   d) 2.0 mil

5. A tape problem that occurs when tape winds unevenly on a reel is called _____.
   a) cupping
   b) drop-out
   c) adhesion
   d) scattered wind

6. In addition to specially designed reels of splicing tape, audio tape editing can be accomplished with _____.
   a) cellophane tape
   b) splicing tabs
   c) masking tape
   d) leader tape

7. In most production situations, which type of edit is preferred?
   a) diagonal cut
   b) vertical cut
   c) horizontal cut
   d) none of the above

8. A razor blade should be replaced during a lengthy editing session as it can become magnetized after a period of time.
   a) true
   b) false

9. If you use splicing tape that is too long _____.
   a) it will be difficult to position it properly
   b) it will create a gap
   c) it will misalign the recording tape
   d) it will overlap the recording tape

10. Which term describes having the end of the material recorded on an audio tape at the outside of the reel?
    a) tails out
    b) tails in
    c) leader out
    d) none of the above

11. Which describes a colored or clear plastic or paper tape that has the same dimensions as audio tape and is often used in audio tape editing?
    a) splicing tape
    b) Scotch® tape
    c) cellophane tape
    d) leader tape

12. Another term for electronic editing is _____.
    a) splicing
    b) dubbing

    c) curling
    d) cupping

13. Chromium-dioxide or chrome equivalent cassette tapes are known as _____ tapes.
    a) Type I
    b) Type II
    c) Type III
    d) Type IV

14. Both splicing and dubbing can be easily accomplished with reel-to-reel, cassette, and cartridge tapes.
    a) true
    b) false

15. Which tape defect is a problem due to poor binding between the plastic base and oxide coating of the audio tape?
    a) print-through
    b) adhesion
    c) cupping
    d) scattered wind

16. Audio tape that is made with an acetate base is often tensilized to prevent stretching problems.
    a) true
    b) false

17. Which type of audio cassette tape requires the highest bias current during recording?
    a) Type I
    b) Type II
    c) Type III
    d) Type IV

18. Audio that is recorded into a digital audio editing system is often labeled as a "region."
    a) true
    b) false

19. Which type of audio tape editing is the most accurate and easiest?
    a) splicing
    b) dubbing
    c) digital audio editing
    d) analog audio editing

20. As a final test on editing, match the terms in the top list (1, 2, 3 . . .) with their definitions in the bottom list (e, r, t . . .), and then select the correct set of answers from the sequences shown in a, b, c, or d below.
    1. _____ print-through
    2. _____ drop-out
    3. _____ Type I
    4. _____ scattered wind
    5. _____ polyester
    6. _____ splicing block
    7. _____ tails out
    8. _____ dubbing
    9. _____ grease pencil
  10. _____ cupping

    e. tape edges turning up
    r. a reel with the end of the tape at the beginning
    t. electronically transferring a signal from one tape to another
    o. a cassette tape that uses ferric oxide as its magnetic material
    c. a crayonlike substance for marking edit points

f. flaking of the oxide coating
u. an uneven wind of tape
p. the plastic base used to make most audio tape
b. magnetic signal on one layer of tape bleeding onto another layer
h. a device that holds tape for editing

a) 1.b 2.f 3.o 4.u 5.p 6.h 7.r 8.t 9.c 10.e
b) 1.f 2.b 3.o 4.e 5.p 6.h 7.r 8.t 9.c 10.u
c) 1.f 2.o 3.b 4.t 5.c 6.e 7.p 8.h 9.u 10.r
d) 1.b 2.f 3.p 4.u 5.o 6.h 7.r 8.t 9.c 10.e

■ ANSWERS

If You Answered A:

1a. Right. It's the bottom layer of the tape sandwich.
2a. Wrong. To tensilize is to prestretch audio tape. (Reread 6.2.)
3a. Right. Polyester tape is preferred because it's strong and little affected by temperature and humidity.
4a. No. Thinner audio tape stretches too easily and is more susceptible to print-through. (Reread 6.3.)
5a. Wrong. This is when the edges of the tape turn up. (Reread 6.5.)
6a. Never. The adhesive material on cellophane tape will bleed through audio tape and gum up the heads of the tape recorder. (Reread 6.8.)
7a. Right. The diagonal cut or 45-degree cutting groove is used in audio tape editing because it gives a smoother sound transition and a stronger edit.
8a. Right. Since the metal razor blade cuts through the magnetic layer of audio tape, it becomes magnetized over a period of time and can degrade the audio signal at the point of the edit if not replaced.
9a. Right. Shorter tape will make for better edits.
10a. Right. The end of an audio tape is called the tail.
11a. Wrong. Splicing tape is used in audio tape editing, but it does not have the same dimensions as audio tape. (Reread 6.8 and 6.12.)
12a. Wrong. Splicing refers to physically cutting the audio tape during the editing process. (Reread 6.7.)
13a. No. These are standard ferric oxide tapes. (Reread 6.4.)
14a. Wrong. For all practical purposes, cassette and cart tapes can only employ dubbing techniques. (Reread 6.7, 6.13, and 6.14.)
15a. No. Print-through is the transfer of the magnetic signal on one layer of tape to the magnetic signal on an adjacent layer. (Reread 6.2 and 6.5.)
16a. You're correct in thinking that to tensilize tape is to prestretch it, but acetate tape is not the type that is tensilized. (Reread 6.2 to find out what tape base is prestretched.)
17a. No. Type I tapes utilize "normal" bias current levels. (Reread 6.4.)
18a. You're not quite right with this answer. While any audio can be marked as a region, we usually think of a region as a portion of a soundfile. A soundfile is the audio that is recorded into a digital editing system. (Reread 6.15.)
19a. No. Splicing can be very accurate when done by a skilled operator, but it really isn't the easiest method. (Reread 6.7 and 6.15.)
20a. Correct. You have finished the exercises for this chapter.

If You Answered B:

1b. Wrong. This is the part that mainly provides strength. (Reread 6.2.)
2b. Wrong. Drop-out is a defect in the oxide coating. (Reread 6.2 and 6.5.)
3b. Wrong. Acetate tape was once favored by broadcasters, but it breaks too often and is susceptible to temperature and humidity problems. (Reread 6.2.)
4b. You're close; 1-mil audio tape may be used by broadcasters, but it's not the best choice. (Reread 6.3.)
5b. Wrong. This is when the oxide flakes. (Reread 6.5.)
6b. Right. Splicing tabs are commercially available precut pieces of splicing tape on applicator strips specially designed for audio tape editing.
7b. No. The 90-degree cut is only used when putting leader tape on an audio tape. (Reread 6.8, 6.10, and 6.12.)
8b. Wrong. (Reread 6.10.)

9b. No. The length of splicing tape has nothing to do with this. (Reread 6.11.)

10b. Wrong. (Reread 6.12.)

11b. No. Scotch® tape should never be used in audio tape editing. (Reread 6.8 and 6.12.)

12b. Right. Dubbing is often considered as electronic editing.

13b. Yes. This is the correct answer.

14b. Right. For all practical purposes, cassette and cart tapes can only employ dubbing techniques.

15b. No. Adhesion is a problem associated with humidity when tape layers stick together. (Reread 6.5.)

16b. Correct. Polyester-based tape is tensilized, not acetate.

17b. No. Type II tapes do utilize a "high" bias current level, but there is a tape type that uses a higher level. (Reread 6.4.)

18b. Yes. What is described is a soundfile, not a region.

19b. Wrong. Dubbing is neither a very accurate nor a very easy method of audio tape editing. (Reread 6.7, 6.13, and 6.15.)

20b. No. You are confused about tape problems. (Reread 6.4.)

If You Answered C:

1c. Wrong. This is where recording occurs. (Reread 6.2.)

2c. Wrong. Adhesion is a problem of tape layers sticking together, usually due to humidity. (Reread 6.2 and 6.5.)

3c. No. Paper tape would most likely be leader tape. (Reread 6.2 and 6.12.)

4c. Yes. Audio tape that is 1.5 mil thick is most often used by broadcasters.

5c. Wrong. Adhesion occurs when tape layers stick to each other, especially after tape has been stored on a reel for some time. (Reread 6.5.)

6c. No. Office tapes should not be used for audio tape editing. (Reread 6.8.)

7c. No. You're way off base with this choice. (Reread 6.8, 6.10, and 6.12.)

9c. No. It shouldn't affect the position of the recording tape. (Reread 6.11.)

10c. Wrong. While there may be leader tape at the end of the audio tape, this is not the correct term. (Reread 6.12.)

11c. Wrong. Cellophane tape (also known as Scotch® tape) should never be used in audio tape editing. (Reread 6.8 and 6.12.)

12c. Wrong. Curling is a problem that can occur when there is a poor binding between the plastic base and the oxide coating of audio tape. (Reread 6.5 and 6.7.)

13c. No. This tape formulation no longer exists. (Reread 6.4.)

15c. Yes. Cupping describes the edges of the audio tape turning up into a cup shape and is caused by poor binding between the tape base and oxide coating.

17c. No. Type III cassette tapes are no longer in use. (Reread 6.4.)

19c. Of course. Digital audio tape editing is extremely accurate and easy to accomplish.

20c. You made many mistakes. (Reread the entire chapter.)

If You Answered D:

1d. No. This is the magnetic layer where recording occurs. (Reread 6.2.)

2d. Right. Print-through occurs when the magnetic signal of one layer of audio tape affects an adjacent layer.

3d. No. This is a metal cassette tape not frequently used in broadcasting. (Reread 6.2 and 6.4.)

4d. No. This is not a standard thickness for audio tape. (Reread 6.3.)

5d. Right. This should be avoided because the edges will be uneven and can be damaged.

6d. No. You're quite confused. Leader tape is not an adhesive tape of any kind. (Reread 6.8 and 6.12.)

7d. No. There is a correct response. (Reread 6.10.)

9d. Wrong. This wouldn't cause the recording tape to overlap. (Reread 6.11.)

10d. No. There is a correct response. (Reread 6.12.)

11d. Right. This accurately describes leader tape.

12d. Wrong. Cupping is a tape defect. (Reread 6.5 and 6.7.)

13d. No. These are metal tapes. (Reread 6.4.)

15d. Wrong. A scatter wind occurs when tape winds unevenly on a reel. Some of the tape edges become exposed as they wind higher or lower than the rest of the tape on the reel. (Reread 6.5.)

17d. Yes. Type IV or metal cassette tapes require the highest bias current during recording.

19d. No. Analog audio editing is what splicing and dubbing techniques are. They are no longer the most accurate or easiest methods for editing. (Reread 6.7 and 6.15.)

20d. No. You are confused about tape construction. (Reread 6.2 and 6.4.)

Projects

■ PROJECT 1

Make two edits in an audio tape on a reel-to-reel recorder.

Purpose

To enable you to feel comfortable editing a vocal audio tape.

Advice, Cautions, and Background

1. If you're not sure of what you're doing, ask the instructor for assistance. Don't take the chance of ruining the equipment.
2. Remember, you are to do two edits, not just one.
3. You'll be judged on the cleanness of your edits, so don't try to edit something that is too tight.

How to Do the Project

1. Familiarize yourself with the operation of the reel-to-reel tape recorder in your production studio. If you have questions, ask your instructor.
2. Assemble the editing tools and supplies that you will need, including splicing block, razor blades, grease pencil, and splicing tape.
3. You can use the material provided on the CD that accompanies this text, or you can record something of your own.
4. Do your edits as follows:
   a. Press the play button and listen to what is recorded.
   b. Select something you wish to edit. Write down on a piece of paper the part you plan to edit with a few words before and after it. Put parentheses around what you plan to take out. For example, for the CD material, you would write: "Today's weather calls for (sunny skies and) a temperature of seventy degrees."
   c. Stop the tape recorder so it's at the exact place you wish to edit—in our example, just in front of "sunny."
   d. You are hearing the tape on the play head (remember ERP). Just right of dead center on the play head, make your edit mark on the audio tape (remember AIM). Be careful *not* to get any grease pencil on the actual tape recorder heads.
   e. Continue playing the tape until you get to the end of your edit—in our example, just before "a."
   f. Using the grease pencil, make your edit mark just as you did before.
   g. Spool out enough tape so that your edit marks can be positioned in the splicing block. Cut the audio tape at the first edit point, according to proper splicing procedure. (You might want to review section 6.10.)
   h. Position the other edit mark in the splicing block, and cut the tape as before. You should now have a loose piece of tape (the unwanted words "sunny skies and") and two pieces of tape with diagonal cuts that are both connected to the two tape reels.
   i. Butt the two tape ends and apply a proper amount of splicing tape. Make sure there are no twists in the audio tape.
   j. Rethread the tape in the recorder, and rewind it a ways by hand. Push play and listen to your edit.
5. Repeat the above steps for your second edit. If using the CD material, you can edit out "seventy-two in Philadelphia" so that the voice says "it will be sixty-six in Chicago and seventy-six in New York."
6. Label the tape with your name and EDITED AUDIO TAPE.
7. Turn in the tape to your instructor to receive credit for the project.

■ PROJECT 2

Dub from one audio tape recorder to another, making two edits as you do so.

Purpose

To enable you to feel comfortable with the type of electronic editing that can be done through dubbing.

Advice, Cautions, and Background

1. If you're not sure of what you're doing, ask the instructor for assistance. Don't take the chance of ruining the equipment.
2. The best way to do this exercise is to use the material provided on the CD that accompanies this text. You could also dub something off the radio.
3. You may have trouble getting the machines to operate so that the edit occurs where you want it to. It's better to make the edit too loose than too tight because one that's loose is easier to correct. You simply need to record again, bringing in the material a little sooner. If your edit is too tight and cuts off the last part of what you had previously laid down, you'll need to lay that part down again.
4. Make sure you have access to two tape recorders and two audio tapes.
5. Remember, you are to do two edits, not just one.

How to Do the Project

1. Record the CD material (or radio material) to a cassette or reel tape. If you record from the radio, make sure you get a commercial and some other voiced material.
2. Connect one tape recorder to another so that the sound can be transferred electronically. You'll need to take a cable from the audio output on the "master" tape recorder to the audio input on the "slave" tape recorder. If your studio is already set up for two tape recorders, you can just go on to the next step.
3. Dub a small portion of what you have on the master tape to the slave machine, just to make sure everything is working correctly. You'll need to put the master tape recorder in play and the slave tape recorder in record. Adjust levels so that both tape machines are about the same. If you have access to "tone," set the machine's VU meters that way. (See the chapter on the audio console.)
4. Play back what you have just recorded and make any necessary adjustments. If everything recorded well, you are ready to start the actual dubbing.
5. Dub what you recorded off the CD or the radio from the master to the slave until you come to a commercial break (on the CD it's the furniture commercial). Hit the pause control on the slave machine right before the commercials begin, and hit pause on the master machine right after the commercials begin. It's better to let both machines run longer than you really want than to cut everything off too short.
6. Play the master tape until you get to the end of the commercial break. Listen carefully so that you know exactly where you want the edit of the program material to begin. Be sure it makes sense, given what was said before the commercial. You may have to edit out some of the intro to the music. Back up at least 5 seconds from your desired edit point, and put the master machine in pause.
7. If you feel confident that the place where you pushed pause on the slave tape is where you want one portion of the program to stop and another to begin, leave it as it is. If you are not sure, listen to the tape again and find the correct point and put the machine in record and pause. Don't go past the point where you recorded, or you'll have an awkward period of no sound at all.
8. Play the master tape, and just as it gets to the spot where you want the edit to be, take the slave tape recorder out of pause. Let both recorders go until the next material you want to edit. For the CD, take out the last spoken sentence—the one that starts, "In other games . . ." and ends, "thirty-four to three."
9. At that point, pause both machines as you did for the previous edit.
10. Go back and listen to the edit you just made on the slave machine. If it is what you want, proceed with the second edit. If not, redo it.
11. To do the second edit, cue the master tape at the end of the sentence about Jeff McKenzie, and cue the slave tape close to where you stopped the recording, just as you did for the first edit.

12. Play the master recorder, and take the slave recorder out of pause, just as you did before. Continue recording to the end of the material; then stop both machines.
13. Listen to the edit you have just finished and make sure it was done correctly.
14. Label the tape with your name and TAPE DUBS.
15. Turn in the "slave tape" to your instructor so that you can get credit for the electronic edits.

■ PROJECT 3

Make several edits utilizing a digital audio tape editor.

Purpose

To enable you to practice editing a vocal audio tape with a digital editing system.

Advice, Cautions, and Background

1. If you're not sure of what you're doing, ask the instructor for assistance. Don't take the chance of ruining the equipment.
2. Remember, you're to do several edits, not just one.
3. You'll be judged on the cleanness of your edits, so be sure you make your edit marks accurately.

How to Do the Project

1. Familiarize yourself with the operation of the digital editing system in your production studio. If you have questions, ask your instructor.
2. You may use the same material used in Project 1, or you may be allowed to record your own. If you can record your own, select some news copy or a weathercast from your news wire service, or write something similar and record it into your editing system. Label this soundfile as EDIT.WAV (or label as required by your system).
4. Do your edits as follows:
    a. Press the play "button," and listen to what is recorded.
    b. Select something you wish to edit. Write down on a piece of paper the part you plan to edit with a few words before and after it. Put parentheses around what you plan to take out. For example: "Today's weather calls for (sunny skies and) a temperature of seventy degrees."
    c. Stop playback so that it's at the exact place you wish to edit—in our example, just in front of "sunny."
    d. Make your beginning edit mark.
    e. Continue playing the audio until you get to the end of your edit—in our example, just before "a."
    f. Make your end edit mark just as you did before.
    g. If you can, preview the edit. You can probably adjust either edit mark so that you have accurately positioned them. If necessary, do so.
    h. Perform the actual edit, and listen to your edit. If you've made some type of error, "undo" the edit, and start it again.
5. Repeat the above steps for another edit. Do a few more similar edits.
6. When you've done several edits, record them all onto a cassette tape.
7. Label the tape with your name and DIGITAL AUDIO EDITING.
8. Turn in the tape to your instructor to receive credit for the project.

CHAPTER SEVEN

Monitor Speakers

Information

7.1 INTRODUCTION

Monitor speakers (see Figure 7.1) are used to listen to the program (and also audition or auxiliary) sound in the radio studio. They are often treated as passive devices that simply exist in the studio, but they are actually quite significant. The sound that comes from them is the final product, so they are important in determining the quality of that product. What you hear on the monitor speakers is an accurate gauge of what the listener will hear.

Speakers are transducers. They work in a manner opposite to that of microphones. Instead of converting sound waves into electrical energy, speakers produce sound from an electrical signal by converting the electrical signal into mechanical energy that produces an audible sound.

7.2 TYPES OF SPEAKERS

The most common type of monitor found in the broadcast studio is a **dynamic speaker**. Also known as an **electromagnetic speaker**, its transducing element, called a "driver," consists of a paper diaphragm, or cone, suspended in a metal frame. Attached at the narrow end of the cone is a voice coil (a cylinder wound with a coil of wire), which is located between powerful circular magnets. When an electrical current is generated in the voice coil, it creates another magnetic force that moves the coil (and cone) in and out, according to the electrical signal entering the coil. The cone vibration causes the surrounding air to move in a like manner, which our ears pick up as sound (see Figure 7.2).

Other speaker drivers (such as **electrostatic**, **ribbon**, or **planar-magnetic)** are considered too exotic (and usually too expensive) for radio use, and you aren't likely to run across them in the production studio.

7.3 BASIC SPEAKER SYSTEM COMPONENTS

The basic components of the typical speaker system (see Figure 7.3) are the woofer, tweeter, crossover, and speaker enclosure. **Woofer** and **tweeter** are names given to drivers or individual speakers used in a speaker system. Since no one speaker design can reproduce the entire frequency range adequately, different speakers were developed to handle different portions of it. A woofer is designed to be able to move the large volume of air that is necessary to reproduce lower frequencies. The cone must be large in size or be able to make large movements, and usually a woofer is a combination of both. However, this bulk prevents the speaker from adequately reproducing the higher frequencies that require rapid cone movement. The tweeter uses a lighter and smaller design; usually a convex dome replaces the cone. There are also midrange speakers, whose names describe the frequencies they reproduce.

An individual speaker is really a speaker system in that most modern speakers use at least a woofer and a tweeter driver. To divide the audio signal and send the proper frequencies to the proper driver, another element of the speaker is used: the **crossover**. A crossover is a network of filters between the input to the speaker and the individual speaker drivers. In the two-way system, one filter would pass all audio below a certain frequency, sending it to the woofer. Another filter would pass all audio above a certain frequency, sending it to the tweeter. Although there is no universal design for the crossover, most dividing points between the bass and treble frequencies are between 500 and 1500 hertz. A speaker that has just a woofer, a tweeter, and a crossover is a **two-way speaker system**. Speakers that employ another driver (such as a midrange) are **three-way systems**.

7.4 SPEAKER SYSTEM DESIGNS

The speaker drivers and crossover are encased in a box (enclosure) that also plays a role in how the speaker sounds. While there are different speaker systems, each is designed to handle the rearward sound wave. Every speaker produces sound both behind and in front: The back sound wave is exactly opposite to the one that goes into the forward listening space. If the two sound waves are allowed to combine naturally, they would be out of phase (see section 7.7) and "cancel" each other out, producing no sound or very diminished sound in a good portion of the frequency range, especially the bass sounds.

FIGURE 7.1 Studio monitor speakers. (Courtesy of JBL Professional)

The two most widely used speaker enclosure designs are the acoustic suspension and bass reflex systems. The **acoustic suspension** (or **sealed box** or **sealed baffle**) design puts the speaker drivers (and crossover) in a tightly sealed enclosure that produces an accurate, natural sound with strong, tight bass. By absorbing it in the enclosure, the acoustic suspension prevents the sound wave generated from the rear of the cone from radiating and disrupting the main sound of the speaker. Acoustic suspension speakers are less efficient than some other designs and require a more powerful amplifier to drive them. The acoustic suspension design also requires a rather large physical enclosure to ensure accurate reproduction of the lowest bass notes.

On the other hand, the **bass reflex** (or **vented box** or **vented baffle**) design is quite efficient and produces a strong bass sound with less power required. The bass reflex speaker enclosure is designed with a port, or duct—an opening that is tuned to allow some of the rear sound (the lower frequencies) to combine in phase and reinforce the main sound from the speaker. Some bass reflex design speakers have been criticized for not having quite as good tonal accuracy as the acoustic suspension design and even adding a boomy quality to the sound. These problems,

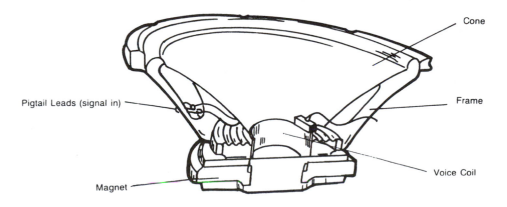

FIGURE 7.2 Dynamic speaker driver.

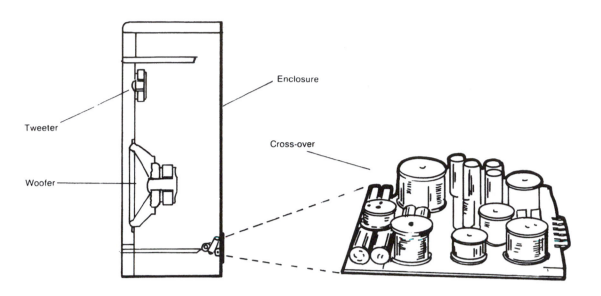

FIGURE 7.3 Basic elements of a speaker system.

however, are often the fault of a particular speaker's tuning and construction, not caused by the bass reflex design. There are many different vented box designs, and most modern bass reflex speakers produce clean, wide-ranging bass.

7.5 SPEAKER SOUND QUALITIES

There's a wide variety of speaker systems to choose from, and, as with some other radio production equipment, the differences between various models may be minimal. One of the important qualities that a good monitor speaker must have is excellent frequency response. We are able to hear sounds or frequencies in the range of 20 hertz to 20 kilohertz, although most of us don't hear quite that low or that high. Top-line broadcast monitors often provide a frequency response range from 35–45 hertz to 18–20 kilohertz. Increased use of CDs and other digital equipment will make speakers that produce as much of this range as possible necessary.

Another quality important for the broadcast monitor speaker is its ability to produce a **flat frequency response**. The speaker should be able to reproduce low, midrange, and high frequencies equally well to produce a natural sound. The speaker itself shouldn't add anything to the audio signal, such as a boosting of the highs.

Perhaps most important is merely how a speaker sounds. Among the combinations of driver types, speaker enclosure design, and crossover frequencies, there is no one speaker configuration that produces the "best sound." It is agreed, however, that a good speaker sound doesn't depend only on the speaker itself. How a speaker sounds is also dependent on the program (what is being played

through it), the dimensions and acoustic properties of the room in which the speakers are heard, the location of the speakers (in relation to the listener), and the listener.

7.6 SPEAKER PLACEMENT

In the radio production room, there may not be many options as to where the monitor speakers are located. Usually they are positioned one on each side of the audio console, but exactly how they are positioned after that is open to various schools of thought. If nothing else, you should try to achieve an acoustically symmetrical layout. Don't put one speaker in the corner next to a glass window and the other speaker in the middle of a wall that's covered with acoustic tiles. That would be a mismatched acoustic environment for the two speakers, rather than a similar one.

One thought is to mount the speakers in the wall. As long as the speakers are isolated from the wall structure—using rubber shock mounts, for example—this flush mount creates an infinite baffle that prevents backwave problems. Large speakers can be built into the wall, providing a loud, clear sound with plenty of "heavy" bass; unfortunately, this type of installation is often not practical for a radio studio.

Some production people feel the ideal sound is obtained when speakers are at or a bit above ear level. This installation is called **near-field monitoring** and can be accomplished by putting small speakers on the audio console "bridge" or on short stands left and right of the console, about 3 feet apart. Keeping the speakers a couple of feet from the back wall will prevent any excessive bass boost. Since the speakers are so close to the listener, you hear mostly direct sound with very little worry about the effect from room acoustics. Toeing in the speakers (angling them toward the listener) or pointing them straight away from the wall will control the treble with speakers pointing toward you providing the greatest high frequency response. Hearing a great amount of clear detail and excellent stereo imaging are characteristics usually associated with near-field monitoring setups. A new concern with monitor speaker placement, especially true of studios that employ near-field monitoring, has to do with the increased use of computer equipment in the broadcast studio. Unless the monitor speakers are magnetically shielded, they must be positioned far enough away from the screen of the computer equipment so that they don't distort the picture.

Perhaps the most practical monitor installation in many production studios is when the speakers are hung from the ceiling or attached to the wall behind the audio console. For the best sound dispersion, the speakers should be hung toward the upper corners of the production room. Keeping them close to the wall prevents a great deal of reflected sound and should produce a full, bass sound at a higher sound level than other possible positions. When speakers are placed closer to a wall, there is more bass sound produced because the usual omnidirectional "soundfield" is cut down, and the sound is concentrated into a smaller dispersion area. Putting speakers totally into the corners of a room (such as the upper corners) may result in too much bass boost, but this is not an uncommon location. Hung speakers also keep counter space available for other production room equipment, and this is an important consideration in most broadcast settings.

Regardless of where the speakers are ultimately placed, the location of the operator in relation to the speakers also plays a role in how they sound, especially with stereo programming. Ideally, the operator is located directly between the two speakers and far enough back from them so that an equilateral triangle is formed if a line were drawn from speaker to speaker and from operator to speaker (see Figure 7.4). If the layout of the production room positions the operator closer to one speaker than the other, the source of all the sound appears to shift toward that one speaker. As a production person, you may not have any control over the speaker placement, but it's important that you realize the effects of speaker placement on the sound you hear.

7.7 PHASE

The concept of **phase** was previously mentioned in the chapter on microphones. Miswiring monitor speakers can also cause phase problems. Each speaker is fed its sound signal from the audio console monitor amp (and sometimes an external amp) by a positive and negative wire. If the wires are reversed on one of the speakers (i.e., the positive wire is connected to the negative terminal), the two speakers will be out of phase. As the driver moves the cone of one speaker in and out, the driver on the other speaker is moving out and in so that the two speaker sounds are fighting each other and tend to cancel out individual sounds, dimin-

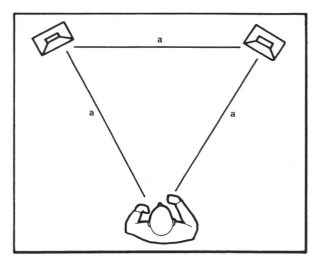

FIGURE 7.4 Speaker-to-listener relationship.

ishing the overall sound quality. This would be especially noticeable if the speakers were reproducing a mono signal. However, since most speakers are wired by the station engineer, phase should not be a problem for the radio production person.

7.8 MONITOR AMPLIFIERS

As you've previously learned, the audio console has an internal monitor amp that provides the signal to drive the monitor speakers. While this is adequate for many studio applications, some production rooms (and control rooms) are set up with external monitor amplifiers. This is merely a more powerful amplifier that provides higher volume levels and clearer reproduction of the sound signal. Remember, the volume of the monitor speakers is only for the pleasure of the operator and has no relationship to the volume of the signal being broadcast or recorded.

7.9 SPEAKER SENSITIVITY

A speaker's sensitivity is the amount of sound (output level) that a speaker can produce from a given input level, much like the sensitivity of a microphone studied in an earlier chapter. It is measured in decibels of sound-pressure level (SPL), and a good quality broadcast monitor will usually have a sensitivity of more than 90 db SPL measured at one meter with 1 watt of input level. There are speakers that range from low-sensitivity to high-sensitivity, but, in reality, this characteristic of a speaker has little bearing on its quality. What it does mean is that a low-sensitivity speaker will need more power to drive it to any given volume level than a high-sensitivity speaker. There is also a relationship between speaker size, sensitivity, and bass response. To increase a speaker's low frequency response, its enclosure can be made larger, or its sensitivity can be decreased. For most broadcast situations, a smaller speaker size offers a lot more flexibility in placement of the speaker and thus is preferred even if it means a low-sensitivity speaker that perhaps requires a bit more powerful monitor amplier.

7.10 HEADPHONES

Headphones are another type of monitor in that they are tiny speakers encased in a headset (see Figure 7.5). Headphones are necessary in radio production because the studio monitor speakers are muted when the microphone is on, and the operator must be able to hear sound sources that are also on. For example, if you're talking over the introduction of a song or reading a commercial over the background of a musical bed, headphones allow you to hear the other sound and the mic sound so that you can balance the two or hit appropriate cues. Headphones are also portable, so sounds can be monitored when an actual monitor speaker might not be available.

Like regular monitor speakers, headphones come in a variety of styles and price ranges. The two main types of headphones found in the production studio are closed headphones and open-air headphones. **Closed headphones**, also known as **circumaural**, have a ring-shaped muff that rests on the head around the ear and not actually on the ear. These headphones are probably the most common in radio since they usually provide a full bass sound and attenuate outside noise better than other styles. However, closed headphones are also heavier and more cumbersome than other styles. **Open-air headphones**, also known as **supra-aural**, have a porous muff, instead of an ear cushion, that rests directly on the ear. Often made of very lightweight material, this design can be very comfortable for the wearer. However, open-air headphones are subject to possible feedback because the audio signal can leak out if driven at high volume levels.

Other headphone types include the tiny **earbud**, which is designed to fit in the ear; **electrostatic headphones**, which are extremely high priced and high quality and require external amplification and special couplers to hook up; and **wireless headphones**, which operate similar to wireless microphones by transmitting an RF, or infrared, audio signal from the source to the headphones.

Unlike consumer headphones, most broadcast-quality headphones are purchased "barefoot"; that is, they have no end connector on them. While most equipment that allows headphone use requires a standard ¼-inch phone connector, there are other situations, and the barefoot headphone allows the engineer to wire them as necessary.

FIGURE 7.5 Broadcast-quality headphones. (Courtesy of AKG Acoustics, Inc.)

Which headphone style is most appropriate is often merely the personal tastes of the announcer using them; however, broadcast-quality headphones should feature large drivers and full (but comfortable) ear cushions and headband. One note of caution for all headphone users: Listening at extremely high volume, especially for extended periods of time, such as in a broadcast situation, can damage your hearing permanently. Good production practice (and common sense) dictates moderate volumes for headphone use.

7.11 CONCLUSIONS

Often, monitor speakers are given little or no thought. Some production people are only concerned with making sure sound comes out of them. But, as you can now understand, there are several variables that can affect how speakers sound, and the role of the monitor speaker in radio production is not as minor as one might initially believe.

Self-Study

■ QUESTIONS

1. A two-way speaker system consists of _____.
   a) a tweeter, a woofer, and a midrange speaker
   b) a tweeter, a woofer, and a crossover
   c) a tweeter and a woofer
   d) a tweeter, a woofer, and two crossovers

2. The transducing element of a speaker is called _____.
   a) a tweeter
   b) a crossover
   c) a woofer
   d) a driver

3. Which speaker enclosure design utilizes a tuned port to provide a highly efficient system with a full bass sound?
   a) acoustic suspension
   b) bass reflex
   c) bass boom
   d) sealed box

4. The individual speaker designed to reproduce higher frequencies is the _____.
   a) woofer
   b) crossover
   c) tweeter
   d) bass reflex

5. For proper stereo sound, the listening angle formed between the speakers and the listener should be 90 degrees.
   a) true
   b) false

6. Good broadcast monitors have a frequency response of about _____.
   a) 20 hertz to 200 hertz
   b) 45 hertz to 2 kilohertz
   c) 20 hertz to 200 kilohertz
   d) 45 hertz to 20 kilohertz

7. The most practical place to locate monitor speakers in a production room is often _____.
   a) near the upper corners, close to the wall
   b) on the counter
   c) not in the room at all, but in an adjoining room
   d) in the middle of the wall, close together

8. If two speakers are out of phase, _____.
   a) the bass sounds will be generated at the rear of the cones
   b) both negative wires will be connected to negative terminals
   c) both positive wires will be connected to positive terminals
   d) the cone of one speaker will be moving out while the cone of the other speaker will be moving in

9. The best type of monitor to use when you need to use a mic in the control room to record a voice over music is _____.
   a) a headset
   b) a tweeter
   c) an acoustic suspension
   d) a bass reflex

10. Which type of monitor speaker will most likely be found in the production studio?
    a) ribbon speaker
    b) electrostatic speaker
    c) dynamic speaker
    d) condenser speaker

11. Which component of a speaker system divides the incoming audio signal into different frequencies and sends the proper frequencies to the appropriate driver?
    a) pigtail leads
    b) tweeter
    c) woofer
    d) crossover

12. Which type of headphone is designed with a porous muff that rests directly on the ear?
    a) closed headphone
    b) circumaural headphone
    c) earbud
    d) supra-aural headphone

13. All speaker enclosures are designed to absorb the backwave produced by the speakers.
    a) true
    b) false

14. Having a high-power, external monitor amplifier in your productions studio will allow you to record or broadcast a louder signal than using the internal monitor amp in the audio console.
    a) true
    b) false

15. As a review of monitors, match the items in the top list (1, 2, 3 . . .) with the choices in the bottom list (s, e, v . . .), and then select the correct set of answers from the sequences shown in a, b, c, or d below.

    1. _____ acoustic suspension
    2. _____ bass reflex
    3. _____ crossover
    4. _____ driver
    5. _____ headphones
    6. _____ tweeter
    7. _____ woofer
    8. _____ circumaural

    s. tiny speakers in a headset
    e. a transducing element for a speaker
    v. a speaker enclosure with a vented port
    t. a tightly sealed speaker enclosure
    l. reproduces lower frequencies
    h. reproduces higher frequencies
    f. sends the proper frequencies to the proper driver
    r. a type of headphone that rests on the head but not actually on the ear

    a) 1.f 2.t 3.s 4.l 5.v 6.f 7.e 8.r
    b) 1.v 2.t 3.f 4.e 5.s 6.l 7.h 8.r
    c) 1.t 2.v 3.f 4.e 5.s 6.h 7.l 8.r
    d) 1.t 2.v 3.f 4.e 5.r 6.h 7.l 8.s

ANSWERS

If You Answered A:
1a. No. This speaker complement would be in a three-way system. (Reread 7.3.)
2a. No. A tweeter is a speaker designed to produce high frequencies. (Reread 7.2 and 7.3.)
3a. No. The acoustic suspension design is relatively inefficient. (Reread 7.4.)
4a. Wrong. The woofer is designed to reproduce the lower frequencies. (Reread 7.3.)
5a. No. This would put the listener directly in front of one of the speakers, and all the sound would appear to be coming out of that speaker. (Reread 7.6; check Figure 7.4.)
6a. No. You're confusing this with the 20-hertz to 20-kilohertz hearing frequencies. (Reread 7.5.)
7a. Correct. This gives the operator good sound and also leaves the counter clear for other equipment.
8a. No. You're confusing this with speaker enclosures. (Reread 7.4 and 7.7.)
9a. Right. A headset is needed to prevent feedback and to hear the music if the speakers are muted when the mic is on.
10a. No. Ribbon speakers are generally too exotic in design and too expensive for broadcast use. (Reread 7.2.)
11a. Wrong. You may be confused because this is a part of an individual speaker that receives an input signal, but after it has been divided into the proper frequencies. (Reread 7.3.)
12a. No. This is a headphone with a ring-shaped ear cushion designed to encircle the ear and rest on the head. (Reread 7.10.)
13a. Wrong. While acoustic suspension or sealed-box designs do absorb the backwave, there are other enclosure designs that utilize the backwave in a different manner. (Reread 7.4.)
14a. You haven't understood the concept of a monitor amp. It has no relation to the broadcast or recorded signal; it only controls the volume of the monitor speakers. (Reread 7.8.)
15a. You made many mistakes. (Reread the entire chapter.)

If You Answered B:
1b. Correct. These are the basic components of a two-way speaker system.
2b. Wrong. The crossover divides the electrical signals and sends them to the speaker drivers. (Reread 7.2 and 7.3.)
3b. Right. This answer is correct.
4b. No. The crossover is not a speaker but an electronic device for sending various frequencies to different speaker drivers. (Reread 7.3.)
5b. Correct. An angle of about sixty degrees should be formed between the listener and the speakers for the best stereo sound.
6b. Wrong. This would only adequately reproduce the lower range of frequencies. (Reread 7.5.)
7b. No. The sound can be good, but the speaker takes up counter space that could be used for something else. (Reread 7.6.)
8b. No. One negative connected to a positive would put them out of phase. (Reread 7.7.)
9b. No. This is only part of a monitor speaker. (Reread 7.3 and 7.10.)
10b. No. Electrostatic speakers are generally too exotic in design and too expensive for broadcast use. (Reread 7.2.)
11b. Wrong. This is a part of a speaker system that reproduces high frequencies. (Reread 7.3.)
12b. No. This is a headphone with a ring-shaped ear cushion designed to encircle the ear and rest on the head. (Reread 7.10.)
13b. Yes. Not all enclosure designs absorb the backwave. For example, bass reflex designs use the backwave to reinforce the main sound.
14b. This is the right response because monitor amps have no relation to the broadcast or recorded signal.
15b. No. You're confused about enclosures and tweeters and woofers. (Reread 7.3 and 7.4.)

If You Answered C:
1c. No. You're close though. (Reread 7.3.)
2c. No. A woofer is a speaker designed to reproduce low frequencies. (Reread 7.2 and 7.3.)
3c. There's no such enclosure design. (Reread 7.4.)
4c. Correct.
6c. No. You're confusing this with the 20-hertz to 20-kilohertz hearing frequencies. (Reread 7.5.)
7c. No. You couldn't hear it if it were in another room. (Reread 7.6.)
8c. No. One positive connected to a negative would put them out of phase. (Reread 7.7.)
9c. No. You're confusing this with enclosure designs. (Reread 7.4 and 7.10.)

10c. Yes. The dynamic speaker is found most often in the production studio.

11c. Wrong. This is a part of a speaker system that reproduces low frequencies. (Reread 7.3.)

12c. No. This is a type of headphone that is designed to fit into the ear. (Reread 7.10.)

15c. You're correct. You have now completed the section on monitor speakers.

If You Answered D:

1d. Wrong. Only a single crossover is required in a speaker system. (Reread 7.3.)

2d. Yes. The driver transforms electrical signals into mechanical energy and thus audible sound.

3d. No. This is another name for an acoustic suspension speaker, which is relatively inefficient. (Reread 7.4.)

4d. No. Bass reflex describes a speaker system design, not an individual speaker. (Reread 7.3 and 7.4.)

6d. Yes. You chose the correct range of frequencies.

7d. Wrong. This would really limit sound dispersion and any stereo imaging. (Reread 7.6.)

8d. Right. The sounds will be fighting each other when this happens.

9d. No. You're confusing this with enclosure designs. (Reread 7.4 and 7.10.)

10d. No. There is no such speaker. (Reread 7.2.)

11d. Correct. A crossover is a network of filters that divides the audio signal into different frequencies and sends it to the proper individual speaker.

12d. Yes. Also known as open-air headphones, supra-aural headphones are designed to rest directly on the ear.

15d. No. You're confused about headphones. (Reread 7.10.)

Projects

■ PROJECT 1

Listen to the sound coming from a speaker (or speakers) from different places in the production room.

Purpose

To make you aware of how sound can change as the relationship between speaker and listener changes.

Advice, Cautions, and Background

1. If your studio is too small for you to hear any differences, just indicate this.
2. The most important thing for your drawings will be to show the relative dimensions of the studio and the position of the speaker(s).

How to Do the Project

1. Make three sketches of your control room, showing where the speaker(s) is (are) located.
2. On the first drawing, put an X where the production person usually sits. On the second drawing, put an X at another spot in the control room where you can stand and listen to the monitors. Do the same for the third drawing.
3. Play some music (you can use the music provided on the CD that accompanies this text) through the monitor speaker(s), and position yourself in each of the three places where you have placed X's. Listen for any differences in the way the music sounds at the three locations.
4. Write a short report detailing how the music sounded at each position.
5. Put your drawings and your report into a packet labeled SPEAKER-LISTENER RELATIONSHIP. Give the packet to your instructor for credit for this project.

■ PROJECT 2

Listen to and compare sound coming from a pair of speakers that are out of phase.

Purpose

To make you aware of how phase can affect the sound speakers reproduce.

Advice, Cautions, and Background

1. Your instructor may decide to present this project as a demonstration to the whole class.
2. Don't rewire the speakers in your studio without approval of the instructor or engineer.
3. You'll need both stereo and monaural music for this project. Both are provided on the CD that accompanies this text. You are advised to use these because mono CDs are hard to find. Some of the early Beatles music was mono, but generally CDs are stereo.

How to Do the Project

1. Set up your studio to play the CD music at a comfortable listening level, and listen to the stereo music first.
2. Pay attention to the sound quality of the disc and its stereo imaging (where the sound appears to be coming from; i.e., left, right, center).
3. Play the mono music and listen as before.
4. Now change the wiring on one speaker only. For example, if a red wire is hooked to a positive speaker terminal (and black to negative), reverse them. Only do this on one speaker so that your studio will be out of phase.
5. Listen to the stereo CD music again. Do you notice any change in the sound quality? You may not hear any noticeable difference.
6. Play the mono CD music again. Do you hear any change in the sound? You probably will hear a noticeable decrease in overall level, bass response, and maybe even some sounds totally canceled out.
7. Rewire the speaker back to its original condition.
8. Write a short report that describes what you heard, and label this report SPEAKER PHASE. Turn this in to the instructor to receive credit for this project.

CHAPTER EIGHT

Connectors, Cables, and Accessories

Information

8.1 INTRODUCTION

The various pieces of equipment used in the radio production studio are all interconnected. This chapter looks at some of the **cables** and **connectors** used for this interconnection. It also discusses some of the accessories that make work in the production studio easier. Several accessories have already been mentioned: audio tape editing supplies, headphones, microphone windscreens, and tape cartridges and cassettes, among others. You might want to review those sections of previous chapters that discuss impedance and mic or line levels before beginning this chapter.

8.2 HARD WIRING AND PATCHING

Audio equipment in the production studio is connected together by two methods: **hard wiring** and **patching**. Hard-wired connections are somewhat permanent (such as the CD player directly connected to the audio console) and are usually soldered or wired by the engineer. Equipment that may, from time to time, be moved from one production area to another (such as the audio tape recorder) is often connected by male and female connectors known as plugs and jacks. More will be said about the typical broadcast connectors in the next few sections of this chapter.

Many pieces of audio equipment and even two different production studios are often connected together through the use of a **patch panel** or **patch bay** (see Figure 8.1), as they are also called. The patch panel is located between the equipment and the audio console so that the input and output of each piece of equipment is wired to the patch panel. In a typical setup, the top row of sockets on the panel is where the audio signal is coming from, and the bottom row of sockets is where the signal is going to. Putting a patch cord into the correct holes in the panel allows you to interconnect and reconfigure the various pieces of equipment or studios. Patch cords are available as either a single-plug cord or a double-plug cord. A double cord works well for stereo patch bays since it has two plugs at each end. One

side of the plug casing will be marked (usually with a serrated edge), so you can keep left and right channels correctly aligned on both the input and output sockets of the patch panel. Of course, single cords work just as well—just remember to always plug in one cord from the top row right channel of the patch panel to the bottom row right channel before plugging in the left channel. That way you won't cross channels as you could if you plugged both left and right channels of the top row before plugging in the bottom row.

Figure 8.2A shows a portion of a patch bay in which a cassette recorder and CD player are linked to two channels on an audio console. With no patch cord put into the panel, CASS 1 is linked to Channel 3 and CD 1 is linked to Channel 4. In patch panels, the top row is internally wired to the bottom row, and the audio signal flows in that manner. When it is unpatched like this, it's known as a "normaled" condition. Figure 8.2B shows another portion of a patch panel that links the output of the audio console to audio recorders. When normaled, the PGM (program) output goes to REEL 1, and the AUD (audition) output goes to DAT 1. The PGM signal could be sent to DAT 1 (as shown) by putting a patch cord from the PGM position on the top row to the DAT 1 position on the bottom row. Now the signal flow has been changed, and that's the purpose of a patch panel—to allow flexibility in configuring the production studio. In this way the audio signals from various pieces of equipment can be sent to the audio console, and the audio console output can be sent to various pieces of equipment.

An alternative to the patch panel is the **audio routing switcher**. The router operates like a patch panel by allowing several input sources to be switched to a single output (or sometimes multiple outputs); however, the switching is done electronically, by selecting the appropriate switches or buttons, rather than using patch cords.

In most broadcast situations, you'll be dealing with four types of connectors: RCA, phone, miniphone, and XLR. In general, the female, or receiving, connectors are called **jacks** and the male connectors are called **plugs**, but often the

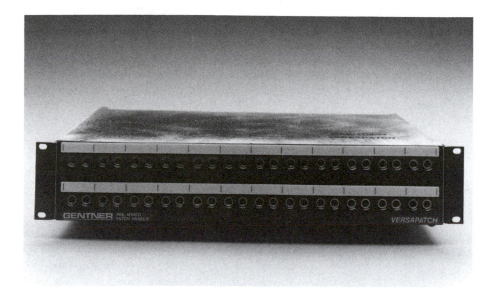

FIGURE 8.1 Patch panel. (Courtesy of Gentner Broadcast Systems)

terms "plugs," "jacks," and "connectors" are used inter-changeably.

8.3 RCA CONNECTORS

The **RCA connector** is also known as the **phono** or **pin connector**. Notice that it is *phono,* not phone. Most home stereo equipment uses this type of connector. In broadcast production, this connector is often used to connect the turntable tone-arm assembly to the turntable preamp, and many CD players utilize RCA connectors. This connector is always a mono connector, so two of them are needed to create stereo. That's why there is a left and right channel output on stereo audio equipment that uses this connection. The male plug consists of a thin outer sleeve and a short center shaft that plugs into the female jack (see Figure 8.3). Although there are female in-line jacks, the female end is most often enclosed in a piece of equipment (such as a tape recorder or preamp) so that the male end will just plug into the equipment. RCA connectors use two-conductor wire that sometimes picks up extraneous electrical noises, such as switches.

8.4 PHONE AND MINIPHONE CONNECTORS

The **phone connector** is also known as the **¼-inch phone**. Notice it is called *phone,* not phono. Most broadcast-quality headphones are connected to the audio console with a phone plug, and most patch bays consist of female phone jacks into which phone plugs are inserted. The **miniphone connector** (also called a **mini**) is most often required to connect portable cassette recorders to other pieces of pro-

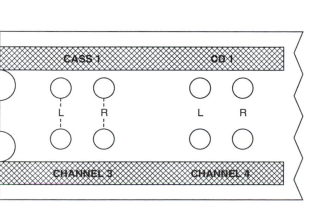

FIGURE 8.2a Portion of a patch panel "normaled."

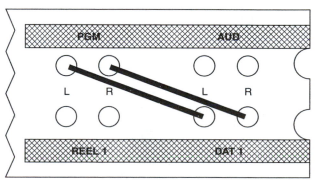

FIGURE 8.2b Portion of a patch panel "patched."

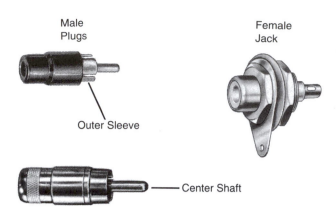

Male
Plugs

Female
Jack

Outer Sleeve

Center Shaft

FIGURE 8.3 RCA connectors. (Courtesy of Switchcraft, Inc., a Raytheon Company)

duction equipment. The output of many portable cassette recorders is a female minijack. As you would expect, the miniphone is a smaller version of the phone connector. While there are various sizes of miniphone connectors, the one used most in broadcast production is 3.5 millimeters.

The phone and miniphone plug consist of a tip and a sleeve, which go into the female jack. Male mono plugs have one insulating ring that separates the tip from the sleeve, and stereo plugs have two insulating rings, which actually define the ring portion of the connector (see Figure 8.4). If the signal is stereo, both the female and male connectors should be stereo.

Again, the female end is often enclosed in a piece of equipment, but you can get in-line jacks if required.

8.5 XLR CONNECTORS

The **XLR connector** is also known as the **Cannon connector** or **three-pin connector**. It's the most common microphone connector in broadcast production use. This connector at-

taches firmly and cannot be disconnected unless the latch lock is pressed. The three prongs of the male plug fit into the conductor inputs of the female jack. The guide pin on the female end fits into the slot for the guide pin on the male end so that the connector can't be put together improperly (see Figure 8.5). Like the RCA connector, the XLR connector is mono so that a stereo connection requires one XLR connector for the right channel and one for the left channel. The three-conductor wiring of the XLR connector makes this a high-quality connection, and it's often found on audio tape cart machines and reel-to-reels in addition to microphones.

8.6 CONNECTOR ADAPTERS

Something else in the connector realm that is very handy to have in a radio production studio is a supply of **connector adapters.** These enable you to change a connector from one form to another.

Let's say, for example, that you need to connect an RCA output to a phone connector input, but the only cable you can find has an RCA connector at both ends. You can convert one of the RCA connectors to a phone connector with an adapter. This is a single piece of metal, which in this case houses a female RCA input at one end and a male phone output at the other (see Figure 8.6). When the male RCA connector is inserted into the female end of the adapter, the signal is transferred from the RCA connector to the phone connector, and from there it can go to the phone input.

As another example, let's say you need a male phone to a male phone connection, but your only cables are male phone to female phone. You can remedy this situation with an adapter that has a male phone plug on each end. Simply place one end of the male adapter into the female jack. Adapters usually come in most handy in emergency situations when some connecting cable fails, so having a variety of them around is good production practice.

Instead of using the connectors previously mentioned, some equipment manufacturers use a nonstandard connec-

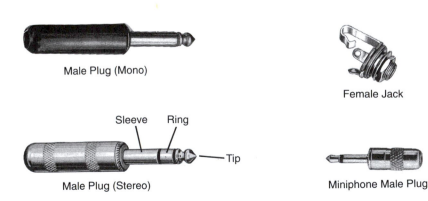

Male Plug (Mono)

Sleeve   Ring

Tip

Male Plug (Stereo)

Female Jack

Miniphone Male Plug

FIGURE 8.4 Phone and miniphone connectors. (Courtesy of Switchcraft, Inc., a Raytheon Company)

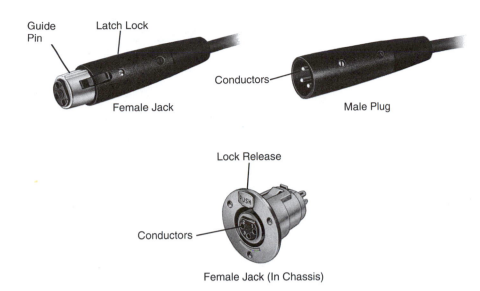

FIGURE 8.5 XLR connectors. (Courtesy of Switchcraft, Inc., a Raytheon Company)

tor for the input and output signals. For example, some tape recorders use a multi-pin connector, similar to the one shown in Figure 8.6, that includes input and output, as well as remote start and cue functions.

8.7 BALANCED AND UNBALANCED LINES

The cable most often used in broadcasting consists of two-stranded-wire conductors that are encased in plastic insulation plus a third uninsulated shield wire, all encased in a foil wrapping and another plastic sheathing (see Figure 8.7). For most wiring practices, the inner wires are designated + (red) and − (black); the uninsulated wire is the ground wire. The audio signal is carried on the positive and negative conductors. This type of cable is referred to as three-wire, or **balanced cable**, and often requires the XLR connector because that is the one designed to connect three wires.

Another type of cable is two-wire, or **unbalanced**. In this configuration, the negative wire also acts as ground. An unbalanced cable is not as good as a balanced cable because unwanted audio interference, such as that created by a nearby electric motor, can creep into the two-wire system. Also, balanced cables can be longer than unbalanced cables without encountering degradation of the signal. Ideally, balanced and unbalanced cable should not be mixed in the same audio setup, but sometimes this can't be avoided because different pieces of equipment are built for different cabling.

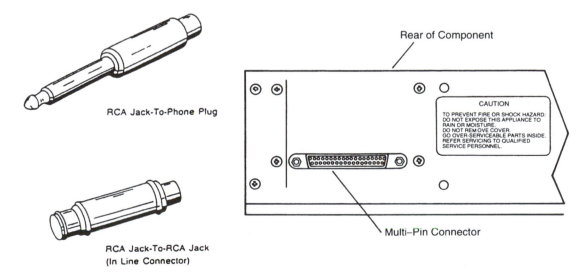

FIGURE 8.6 Connector adapters and multi-pin connector.

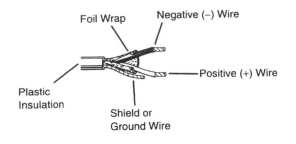

FIGURE 8.7 Typical broadcast audio cable. (Courtesy of Cooper Industries—Belden Division)

8.8 IMPEDANCE

Impedance is a complex way of looking at the total resistance in an audio circuit. The impedance of a particular piece of audio equipment is not obvious, but it can be either high or low. Most broadcast equipment is designed for low impedance because such a design gives less resistance to the signal. In general, a low impedance audio source operates into a low- or high-impedance input (although the low output/high input mismatch may not give the best-quality results), but a high impedance source (output) should not be placed into a low-impedance input because it will cause sound distortion. There are transformers that will match a high impedance to a low impedance and vice versa. But again, the best production practice is not to mix high and low impedances.

8.9 MIC, LINE, AND SPEAKER LEVELS

Equipment inputs and outputs can be one of three levels: mic, line, or speaker. You can think of these levels as very low for mic level (which usually must be preamplified to be used further), normal for line level (most equipment will use line levels), and very high for **speaker levels** (designed to drive a speaker only). Problems arise when various levels are mismatched. For example, if you tried to feed an audio tape recorder from a speaker-level source, you would probably distort the recording because the speaker-level source is too loud, and there is no control to turn it down. Another problem would occur if you fed a mic-level signal into a line-level input. In this case, the signal would be too low to be usable because mic levels must be preamplified to a usable level. Most broadcast equipment inputs and outputs are clearly designated as mic, line, or speaker level, and good production practice sends only the proper output to the proper input.

Fortunately, most of the cabling of production room equipment has been done by the engineer, and all the connections of various pieces of equipment have been worked out so that everything matches.

FIGURE 8.8 Broadcast studio timer. (Courtesy of Radio Systems, Inc.)

8.10 STUDIO TIMERS

Timers have already been mentioned because some audio consoles have built-in timers, but it is not uncommon to find a separate studio timer in the radio production room (see Figure 8.8). Since the timing of radio production work is so important, an accurate timing device is crucial. Most studio timers are digital, showing minutes and seconds, and include at least start, stop, and reset controls. Many timers can be interfaced with other equipment (such as tape recorders and turntables) so that they automatically reset to zero when that piece of equipment is started. Shorter timers (10 minutes) are usually adequate for radio production work, and 24-hour timers are often found in the studio.

8.11 TAPE DEGAUSSERS

The audio tape degausser, or **bulk eraser**, is merely a strong electromagnet used to erase audio tape in the production studio. If you remember the information in the audio tape recorder chapter, you know audio tape cartridges must be bulk erased, and it's good production practice to erase all audio tapes, even though reel-to-reel and cassette tape machines have erase heads in them. Some bulk erasers are designed for tabletop use (see Figure 8.9), while others can be handheld.

Erasing tapes is a simple operation, but it's often done improperly by beginning production people and studio pros alike. To erase an audio tape (reel-to-reel), do the following:

a. Extend the tape to be erased an arm's length away from the degausser and then turn the degausser on. (If the tape is already sitting on the degausser when you turn it on, the transient surge of turning the unit on will put a click on the audio tape that can't be erased by normal operation of the eraser).

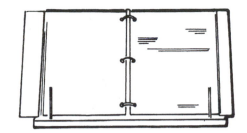

FIGURE 8.10 Copy holder.

8.12 COPY HOLDER

A **copy holder** is an often overlooked, but highly useful, radio production room accessory (see Figure 8.10). It's merely a small easel that can sit on the audio console. It usually has its own stand that's designed to hold a piece of paper (radio script) in front of you. Once you're doing production work, you'll see how important this is. Not only does it free your hands from holding the script (as you often have to be turning a pot or throwing a switch with your hands as you're reading the script), but it positions the script where it should be for proper mic technique. Often broadcasters merely set the script on the counter in front of the audio console and read it from there; however, this draws the head down toward the counter and the script and away from speaking into the microphone. If your production facility doesn't have a copy holder, you may want to suggest purchasing one.

8.13 CLEANING SUPPLIES

Cleaning supplies are common broadcast accessories. The basic supplies are cotton swabs and head cleaner, used to keep audio tape recorders clean. The swabs are not Q-Tips, but long wooden sticks with a cotton tip on one end. These are excellent for getting at hard-to-reach tape heads and removing edit pencil grease, tape residue, and other dirt. Expensive head-cleaning solvents can be replaced by denatured alcohol, but avoid using lower-quality isopropyl, or rubbing alcohol, because of its water content.

Record-cleaning and stylus-care systems are also useful aids in keeping the tools of audio production clean, but the newest cleaning systems have been developed due to the increased use of CDs. The Discwasher® CD Hydrobath (see Figure 8.11) cleans CDs of dust, fingerprints, and other forms of dirt using hydrodynamic principles. The CD and cleaning solution are put in the cleaner/dryer case. In less than a minute, as the CD spins at high speed, the cleaning solution is directed on the CD surface, dirt is trapped and removed, and the CD is spun in the opposite direction to dry it. This type of "no contact" cleaning prevents any accidental damage to the disc surface that could happen with hand cleaning.

FIGURE 8.9 Bulk eraser (degausser). (Courtesy of Audiolab Electronics, Inc.)

b. Place the tape on the guide pin, and rotate it two or three revolutions.

c. Turn the tape over, and rotate it another two or three times.

d. Move the tape an arm's length away from the degausser, and turn it off.

Cartridge and cassette tapes are erased in a similar manner. Using a handheld degausser is similar, except that the tape is held still, and the degausser is moved. The important points to remember are to turn the degausser on and off with the tape away from it and to keep the degausser turned on as you turn the audio tape from one side to the other. Since the degausser is an electromagnet, it's a good idea to keep your watch out of close contact with it when erasing tapes. Also, keep tapes you don't want erased away from the degausser when it is operating. Once a tape is erased, it can't be restored except by rerecording the material.

There is another type of cartridge eraser that resembles a cartridge machine in appearance. When a cart is inserted and the eraser is turned on, the electromagnet comes on gradually (to avoid a surge that would cause a click on the audio tape) and thoroughly erases the cartridge. This type of unit (called an **eraser/splice finder**) also finds the splice on the audio tape cartridge. As the tape runs in the machine, a sensor locates the splice (where the tape was joined together to form the endless loop) and stops the tape just past the splice, which is the point where you would begin to record. This is important because you do not want to record over the splice.

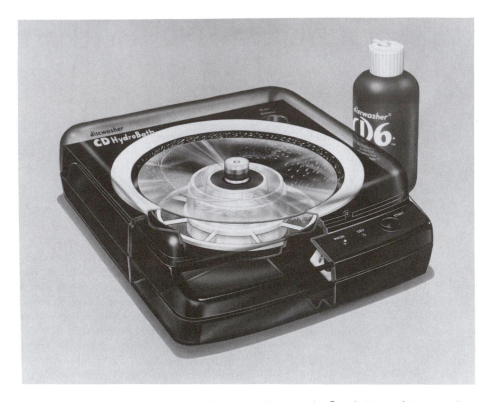

FIGURE 8.11 CD cleaning system. (Courtesy of Discwasher®, a division of Recoton Corporation LIC, New York)

With more and more computer equipment making its way into the radio production studio, cleaning supplies for the monitor screen and keyboard need to become an added broadcast accessory.

8.14 JEWEL BOXES AND RECORD SLEEVES

Broadcast suppliers offer empty plastic cases for CDs, known as **jewel boxes**. These are needed because many promotional CDs come to radio stations in paper jackets rather than cases. Good production practice dictates keeping the CD in its case when not in use to prevent scratches or problems from dust and dirt.

Similarly, **record sleeves**, or **shucks**, are heavy paper jackets that are used to replace the flimsy record jackets that come with 45 RPM records. Not only do they protect the record better, but they're available in several colors so that records can be easily categorized.

8.15 TELEPHONE INTERFACE

The **telephone interface**, or **telephone coupler** (see Figure 8.12), is a piece of equipment designed to connect telephone lines to broadcast equipment. In a basic configuration, the telephone goes through the interface and comes into the audio console on its own channel (just like a CD player, for example). The caller volume is controlled with that channel's fader, in addition to a caller volume control on the interface. Once a call is taken by the announcer and the interface is switched on, the announcer talks to the caller through the studio microphone, not the telephone instrument (which can be hung up at this point) and hears the caller through the headphones.

The interface electronically maintains an isolation between the studio "send" signal and the caller "return" signal, providing a high-quality, clear telephone signal. In addition, many radio stations have installed high-quality digital (ISDN) services that give a clearer signal than the older analog systems. Since the telephone becomes just another audio input, it can be easily recorded (and later edited), mixed with other sound sources, and otherwise manipulated for production use.

8.16 CONCLUSIONS

Not every accessory used by the professional radio broadcaster has been mentioned in this chapter, but you have been introduced to the most common items used in the production studio, and you're less likely to run across something that makes you ask, "What's this for?"

FIGURE 8.12 Telephone interface. (Courtesy of Gentner Broadcast Systems)

Self-Study

■ QUESTIONS

1. When a patch panel is normaled, a patch cord is used to link broadcast equipment assigned to the top row of the panel to the equipment assigned to the bottom row.
   a) true
   b) false

2. Which broadcast connector has a guide pin?
   a) RCA
   b) phone
   c) XLR
   d) phono

3. Which broadcast connector has a sleeve and tip?
   a) RCA
   b) Cannon
   c) XLR
   d) phone

4. Which connector is most likely to be used for a patch bay?
   a) phone
   b) miniphone
   c) RCA
   d) multi-pin connector

5. Which connector is always mono?
   a) phone
   b) miniphone
   c) ¼-inch phone
   d) RCA

6. A connector adapter is used _____.
   a) to transfer a signal in a patch bay
   b) to change a connector from one form to another
   c) to make a balanced line unbalanced
   d) to change an audio signal from high to low impedance

7. A balanced cable usually has _____.
   a) two wires
   b) three wires
   c) high impedance
   d) two ground wires

8. Unbalanced cables are more susceptible to interference than balanced cables.
   a) true
   b) false

9. The normal inputs of an audio tape recorder are designed for which level of audio signal?
   a) mic
   b) line
   c) speaker
   d) none of the above

10. Which production room accessory is used to erase audio tapes?
   a) jewel box
   b) Cannon connector
   c) impedance transformer
   d) degausser

11. Which production room accessory is used to protect and help categorize records?
   a) Discwasher
   b) timer
   c) record shucks
   d) jewel box

12. When using a degausser, you should turn it on while holding the tape an arm's length away from it so that _____.
   a) you give the degausser time to come up to full power
   b) you don't demagnetize your watch
   c) you don't erase tapes that are in the vicinity
   d) you don't put a click on the tape

13. One of the advantages of a copy holder is _____.
   a) it frees your hands
   b) it magnifies the script
   c) it moves your head away from the mic
   d) it makes a convenient place to keep production room cleaning supplies

14. Which production room accessory is used to connect telephone lines directly to broadcast equipment?
   a) audio routing switcher
   b) telephone coupler
   c) patch panel
   d) XLR connector

15. As a review of this chapter, match the nine pictured items (1, 2, 3 . . .) with their names (p, x, m . . .), and then select the correct set of answers from the sequences shown in a, b, c, or d below.
   p. phone connector
   x. XLR connector
   m. mini connector
   r. RCA connector
   c. copy holder
   d. degausser
   a. patch panel
   h. phone coupler
   t. studio timer

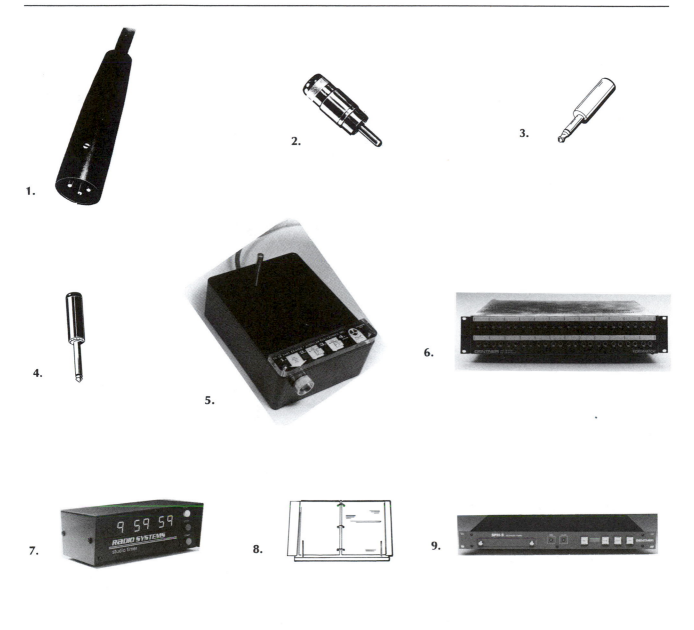

a) 1.r 2.m 3.p 4.x 5.d 6.a 7.t 8.c 9.h
b) 1.r 2.x 3.p 4.m 5.a 6.c 7.t 8.h 9.d
c) 1.x 2.r 3.p 4.m 5.d 6.h 7.t 8.c 9.a
d) 1.x 2.r 3.m 4.p 5.d 6.a 7.t 8.c 9.h

■ ANSWERS

If You Answered A:

1a. Wrong. A normaled patch panel is actually unpatched. (Reread 8.2.)

2a. No. (Review Figures 8.3–8.5; reread 8.3–8.5.)

3a. No. The phono connector has an outer sleeve but no tip. (Review Figures 8.3–8.5; reread 8.3–8.5.)

4a. Right. A male phone plug is the most likely connector to use with a patch bay.

5a. No. The ring on a phone connector makes it stereo. (Review Figures 8.3–8.5; reread 8.3–8.5.)

6a. No. You would be very unlikely to use an adapter with a patch bay. (Reread 8.2 and 8.6.)

7a. No. You have it confused. (Reread 8.7.)

8a. Right. Because one wire conducts the signal and also acts as ground, they are more likely to pick up unwanted noise.

9a. Wrong. A mic level might be fed into a tape recorder, but it must be preamplified to be usable. (Reread 8.9.)

10a. Wrong. This is used to store compact discs. (Reread 8.11 and 8.14.)

11a. No. A record-cleaning system like the Discwasher helps protect records, but it has nothing to do with categorizing them. (Reread 8.13 and 8.14.)

12a. Wrong. Bulk erasers are at full power as soon as you turn them on. (Reread 8.11.)

13a. Yes. In this way you are freer to operate the various equipment controls.

14a. Wrong. An audio router can be involved in selecting various inputs and outputs, but it can't connect a telephone line by itself. (Reread 8.2 and 8.15.)

15a. No. You made many mistakes regarding connectors. (Reread 8.3–8.5, and study the accompanying figures.)

If You Answered B:

1b. Yes. When a patch cord is put into a patch panel, it is no longer normaled.

2b. No. (Review Figures 8.3–8.5; reread 8.3–8.5.)

3b. Wrong. Cannon is just another name for the XLR connector. (Review Figures 8.3–8.5; reread 8.3–8.5.)

4b. No. You're warm but not correct. (Reread 8.2 and 8.4.)

5b. No. The ring on the miniphone connector makes it stereo. (Review Figures 8.3–8.5; reread 8.3–8.5.)

6b. Correct. It transfers the signal so that another form of connector can be used.

7b. Right. There are three wires.

8b. Wrong. (Reread 8.7.)

9b. Right. Line-level inputs and outputs are standard in broadcast production.

10b. No. This is a common broadcast connector. (Reread 8.5 and 8.11.)

11b. No. This has nothing to do with records. (Reread 8.10 and 8.14.)

12b. No. You take the risk of demagnetizing your watch regardless of how you turn it on. (Reread 8.11.)

13b. No. (Reread 8.12.)

14b. Correct. This is the best answer.

15b. You made many mistakes. (Reread the entire chapter.)

If You Answered C:

2c. Right. The XLR jack has a guide pin that prevents it from being connected incorrectly.

3c. No. (Review Figures 8.3–8.5; reread 8.3–8.5.)

4c. No. (Reread 8.2, 8.3, and 8.4.)

5c. No. A $1/4$-inch phone or phone connector can be either stereo or mono. (Review Figures 8.3–8.5; reread 8.3–8.5.)

6c. No. Adapters are not related to balance. (Reread 8.6 and 8.7.)

7c. No. If anything, balanced cable usually goes with low impedance, but the two are not necessarily connected. (Reread 8.7 and 8.8.)

9c. No. Speaker level is quite high and will usually distort if fed to the input of a recorder. (Reread 8.9.)

10c. No. This is used to match high- and low-impedance equipment. (Reread 8.8 and 8.11.)

11c. Yes. Record shucks are color-coded protective sleeves.

12c. No. Nearby tapes can be erased regardless of how you turn it on. (Reread 8.11.)

13c. No. If anything, it positions your head so that you have proper mic technique. (Reread 8.12.)

14c. Wrong. A patch panel could be involved in wiring a telephone line to an audio console, but it can't connect a telephone line by itself. (Reread 8.2 and 8.15.)

15c. No. You confused several things. (Reread 8.2, 8.4, and 8.15.)

If You Answered D:

2d. No. Phono is just another name for the RCA connector. (Review Figures 8.3–8.5; reread 8.3–8.5.)

3d. Correct. The miniphone connectors also have them.

4d. No. (Reread 8.2, 8.4, and 8.6.)

5d. Correct. You must use two RCA connectors, one for each stereo channel.

6d. No. There are devices to do this, but they are transformers, not connector adapters. (Reread 8.6 and 8.8.)

7d. Wrong. (Reread 8.7.)

9d. No. There is a correct answer. (Reread 8.9.)

10d. Correct. The degausser erases audio tapes.

11d. No. This is the plastic case that protects a CD. (Reread 8.14.)

12d. Correct. Bring the tape in gradually to the degausser.

13d. No. You're getting two different production room accessories confused. (Reread 8.12 and 8.13.)

14d. Wrong. You're quite confused if you chose this answer. (Reread 8.5 and 8.15.)

15d. Correct. You have now completed this chapter.

Projects

■ PROJECT 1

Inventory the broadcast accessories found in your production room or studio.

Purpose

To familiarize you with available accessory items.

Advice, Cautions, and Background

1. Make a form similar to the one on the following page to conduct your inventory.
2. You can use items in any production room or studio available at your facility. You'll be asked to note the locations.

How to Do the Project

1. List the brand name and model of *each item,* if possible.
2. Make sure you record the location of each item (e.g., Production Studio A, PDX 1).
3. Fill in all specifics on the form.
4. If your production facility doesn't have a particular item, indicate this on the form.
5. Turn in your inventory form to the instructor for credit for this project. Don't forget to put your name on it.

BROADCAST ACCESSORIES INVENTORY FORM

1. Item: Degausser   Location:
   Brand Name & Model #:
   [ ] Handheld
   [ ] Tabletop

2. Item: Studio timer   Location:
   Brand Name & Model #:
   Maximum length of time:

3. Item: Cart rack   Location:
   Brand Name & Model #:
   [ ] Wall mount
   [ ] Rack mount
   [ ] Lazy Susan

4. Item: CD/Record cleaning system   Location:
   Brand Name & Model #:

5. Item: Audio tape cartridge   Location:
   Brand Name & Model #:
   Length:

6. Item: Microphone windscreen   Location:
   Brand Name & Model #:

7. Item: Copy holder   Location:
   Brand Name & Model #:

8. Item: Telephone Interface   Location:
   Brand Name & Model #:

9. List additional items in a similar fashion.

■ PROJECT 2

Degauss a tape.

Purpose

To make sure you know the proper procedure for bulk erasing audio tapes.

Advice, Cautions, and Background

1. This exercise is designed so that you can use either a table-mounted or handheld degausser. If your facility has only a cartridge eraser, use that and modify the project to fit the operation of that type of degausser.
2. Remember to remove your watch, or be careful with it around the eraser so that it doesn't become demagnetized.
3. Play the tape both before and after the degaussing.

How to Do the Project

1. Find a cassette or reel-to-reel tape that has material on it that's no longer needed. (Your instructor may provide you with one.)
2. Play a bit of the tape so that you're sure something is on it.
3. Locate your facility's degausser.
4. Place the tape about an arm's length from the degausser, and turn it on.
5. Gradually bring the tape to the degausser if you're using a table-mounted model, or bring the degausser to the tape if you're using the handheld model.
6. For the table model, place the tape on or near the guide pin, and rotate it two or three revolutions. For the hand-held model, move the degausser over the entire tape two or three times.
7. Turn the tape over. (Do *not* turn the degausser off.)
8. Repeat step 6 for the other side of the tape.
9. Move the tape an arm's length from the degausser, and turn it off.
10. Play part of the tape to make sure there are no clicks or remaining sounds.
11. Turn in the erased tape to your instructor for credit for this project.

CHAPTER NINE

Signal Processing Equipment

Information

9.1 INTRODUCTION

Signal processing is nothing more than altering how something, such as an announcer's voice or a CD, sounds. We've already seen some forms of audio processing in the equalization capabilities of some audio consoles and the bass roll-off of some microphones, but most signal processing is accomplished by the use of separate components. Some digital equipment, especially digital audio workstations, also includes features that allow for signal processing the audio signal.

Usually signal processing involves the manipulation of the frequency response, imaging, or dynamic range of the sound signal. Frequency response refers to the range of all frequencies (pitches) that an audio component can reproduce and we can hear, i.e., around 20 hertz to 20,000 hertz. **Imaging** refers to the perceived space between and behind monitor speakers and how we hear individual sounds within that plane. Dynamic range refers to the audible distance between the softest sounds that can be heard over noise and the loudest sounds that can be produced before distortion is heard.

Several signal-processing devices will refer to the terms "wet" and "dry." The incoming, or unprocessed, audio signal is considered a *dry* signal, and the outgoing audio signal that has been processed is considered the *wet* signal.

This chapter focuses on those electronic processors commonly used in radio; the amount of equipment available for signal processing in any one production facility, however, can vary widely, from essentially none to a veritable smorgasbord of electronic black boxes.

9.2 EQUALIZERS

The most commonly used signal processor is the equalizer. An equalizer allows for manipulation of the frequency response by adjustment of the volume of selected frequencies and can be thought of as a fancy **tone control**. You're familiar with bass and treble tone controls—most home stereo systems have them. When you turn up the treble control, you increase the volume of the higher frequencies. Unfortunately, a treble control turns up all the higher frequencies.

An equalizer, on the other hand, offers greater flexibility and allows the operator to differentiate, for example, between lower-high frequencies and upper-high frequencies and make different adjustments to each.

Most equalizers found in the radio production studio are **unity-gain** devices. The sound signal that has passed through the equalizer's circuits is no stronger or weaker overall than the input sound signal was. In other words, if the processor's controls are set at zero they should have no processing effect on the in-coming audio signal. Of course, if the signal is processed by increasing the level of several different frequencies, then the overall volume of the outgoing signal will be louder than the unprocessed signal.

9.3 THE GRAPHIC EQUALIZER

The two main kinds of equalizers found in radio production are the graphic equalizer and the parametric equalizer. The **graphic equalizer** is more common and derives its name from the rough graph of a sound's altered frequency response formed by the slider control settings on the equalizer's front faceplate (see Figure 9.1).

A graphic equalizer divides the frequency response range into separate bands, usually at one-third-, one-half-, or full-octave intervals. If the first band of a full-octave equalizer was at 60 hertz, the second would be at or near 120 hertz, the third at 240 hertz, and so on. Equalizers range from a mere 5 or 6 bands to over 30 bands. Obviously, the more bands you have to work with, the greater control you have, but it also becomes harder to correctly manipulate the equalizer. Most broadcast-quality graphic equalizers are full-octave, ten-band equalizers. Each band has a slider volume control that's off in a middle or "flat" position and can move up to increase ("boost"), or down to attenuate ("cut"), the volume at that particular frequency. The volume range varies, but +12 to –12, or twelve decibels of boost or cut, is a common configuration.

Figure 9.2 shows a basic five-band equalizer. Note how the slider setting would demark a frequency response graph, indicating how the original sound was being manipulated. If we were playing a piece of music through this equalizer, the increase setting at 60 hertz would give the

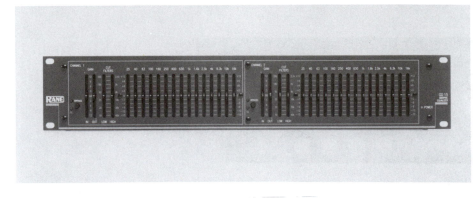

FIGURE 9.1   Graphic equalizer. (Courtesy of Rane Corporation)

drums extra punch; cutting back at 250 hertz would help minimize bass boom; the increase at 1 kilohertz would add brilliance to the voices; the decrease at 4 kilohertz would minimize the harshness of the sound; and the increase at 8 kilohertz would add presence to the highs.

There is no one correct setting for an equalizer, and we could just as easily have had all different settings yet produced an excellent sound. There are, however, settings that would make the sound very poor, and the production operator needs to beware of altering the sound too much.

9.4   THE PARAMETRIC EQUALIZER

The **parametric equalizer** gives the operator even greater control over the sound because it allows not only volume control of specified frequencies, but control over the actual center frequency and bandwidth selected. For example, the five-band graphic equalizer mentioned above had a set center frequency band at 1 kilohertz, but a parametric equalizer allows us to choose an exact frequency (instead of 1000 hertz, maybe 925 hertz or 1200 hertz). If the graphic equalizer were increased at 1 kilohertz, not only would that frequency get a boost, but so would the adjacent frequencies

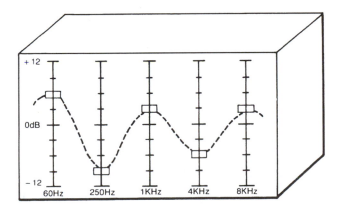

FIGURE 9.2   A simple five-band graphic equalizer.

(perhaps from 500 to 1500 hertz), according to a preset bandwidth determined by the manufacturer. Many parametric equalizers also allow the operator to adjust that bandwidth. For example, still using the 1-kilohertz center frequency, the parametric equalizer operator could select a bandwidth of 800–1200 hertz or a narrower bandwidth of 950–1050 hertz to be equalized. Most equalizers also have a switch that allows the sound signal to pass through the equalizer unaffected, or unequalized. Figure 9.3 shows one type of parametric equalizer.

9.5   EQUALIZER USES

We've already said that the general use of an equalizer is to alter or change the sound character of an audio signal. The term **EQ** refers to the general process of equalization when an equalizer is used to cut or boost the signal level of specific frequencies. Specifically, equalizers are often used to cut down on various forms of audio noise, to alter recordings to suit individual taste, and to create a special effect.

EQ is often used to deal with hiss, one of the common forms of noise in production work. **Hiss** describes a high-frequency noise problem inherent in the recording process. Hiss sounds exactly as you would imagine and is heard by the human ear within the frequency range of 2–8 kilohertz. Using an equalizer to cut or turn down the frequency settings closest to this range can attenuate hiss to a less-noticeable level. However, even with careful adjustment, it's often impossible to eliminate noise entirely. Remember, you also affect the program signal as you equalize, so EQ is usually a compromise between less noise and a still-discernible signal.

EQ is sometimes used to combat hum, which is another form of noise in production work. **Hum** is a low-frequency problem associated with leakage of the 60 hertz AC electrical current into the audio signal. A poor or broken ground in any part of the electrical circuit can cause hum, and it, too, sounds exactly as you would imagine. Attenuating frequencies right around 60 Hz would lessen the problem.

The equalizer may also be used for altering the sound of

FIGURE 9.3 Parametric equalizer. (Courtesy of Orban Associates, Inc.)

CDs (or actually any music) to suit an individual taste. For some production work, you may need to accent the bass in a particular piece of background music and boosting frequency settings around 50 Hz certainly could help do so. Equalizing anything is very subjective, and what sounds good to one listener may not to another. The key is to use any effect moderately, and experiment to find just the right sound.

Sometimes equalization is undertaken to achieve a special effect. For example, cutting down most of the lower frequencies leaves a very tinny-sounding voice that might be perfect as a robot voice for a particular commercial or radio drama. You could produce an effect that approximates an old-fashioned, poor-quality telephone sound by boosting the frequency settings from about 600–800 Hz and cutting the frequency settings on either side of this range. Many such effects are achieved by experimenting with various settings of the equalizer.

9.6 FILTERS

Filters are less likely to be found in the radio production room, but since they are a type of equalizer, a short discussion of them may prove useful. Instead of manipulating a specific frequency, filters affect a whole range of the audio signal. For example, a **low cut filter** cuts, or eliminates, frequencies below a certain point, say 500 hertz. Instead of a normal frequency response of 20–20,000 hertz, once the signal goes through this filter, its response is 500–20,000 hertz. A **low pass filter** works in the same fashion but lets frequencies below a certain point pass or remain unaffected. In other words, a cut filter and a pass filter are opposites in their action. A **band pass filter** cuts frequencies except for a specified band. It has a low cut point (maybe 1,000 hertz) and a high cut point (maybe 3,000 hertz). When a signal is sent through this filter, only that portion of the signal between the two cut points is heard. A **band reject** (or **cut**) **filter** is just the opposite of the band pass filter in that it allows frequencies to pass except for a specified frequency

range. A **notch filter** is a special filter that completely eliminates an extremely narrow range of frequencies or one individual frequency.

Usually filters are used to correct a specific problem. For example, a record may have a scratch that shows up around 11 kilohertz, and a notch filter is used to eliminate that frequency. Of course, filters eliminate both the problem *and* the actual program signal, so careful use is necessary to maintain a good audio signal after filtering has taken place.

9.7 NOISE REDUCTION

We've mentioned the problem of noise inherent in the production process. Signal-processing devices, known as noise-reduction systems, have been devised to help prevent noise. Note that these electronic devices can't get rid of noise that already exists; their job is to prevent noise from being added to a recording. Digital recording does not need to use noise-reduction devices because the noises they are meant to eliminate are ones inherent in the analog recording process. There are two systems commonly used with analog systems: Dolby and dbx. Both are **companders**— their general operation is to compress (reduce the dynamic range of) an audio signal during recording and then expand the signal during playback—but they are not compatible systems.

9.8 DOLBY NOISE-REDUCTION SYSTEMS

There are several **Dolby** noise-reduction systems in use in recording studios, broadcast stations, home stereo equipment, and movie theaters. The Dolby B and C systems and Dolby S and SR are most likely to be found in the production studio, often built into the electronics of various audio tape recorders.

The two-part Dolby process consists of increasing the volume of the program signal at certain frequencies (specifically at the upper end of the frequency spectrum, where the signal is most likely to be lost to noise) before the recording begins. If the particular frequencies aren't affected by noise, they aren't altered by the Dolby encoding process. During playback, levels that were increased are decreased, but noise in the recording process was not boosted, so it seems lower in relation to the program level. Depending on the system employed, 10–25 decibels of noise reduction can be attained using the Dolby process. As with all other signal-processing equipment, careful use is required to achieve the results the operator wants. Setting proper levels prior to recording with Dolby is important so that both encoding (recording) and decoding (playback) of the audio signal are at the same level.

Dolby SR (for spectral recording) and Dolby S are the newest and most sophisticated noise-reduction systems from the Dolby Laboratories. Dolby S (derived from the SR system) is designed to be an improved version of the Dolby B and Dolby C systems.

9.9 DBX® NOISE-REDUCTION SYSTEMS

There are two **dbx**® noise-reduction systems: Type I and Type II. Type II is most often found in broadcasting (see Figure 9.4), and there is no compatibility between the two dbx systems. Unlike Dolby, dbx compresses the signal over the entire frequency range during recording. Again, since the signal goes through the noise-reduction unit before recording, is processed and recorded, and then is played back through the noise-reduction unit, inherent recording noise is mostly covered up.

Noise reduction of 40 decibels can be attained using the dbx system. The audio signal is compressed during recording by a two-to-one ratio (in other words, the dynamic range is cut in half), with the loud levels greatly reduced and the soft levels boosted by a carefully designed frequency response preemphasis. During playback, the signal is expanded by a one-to-two ratio, with deemphasis, so that the original dynamic range is restored. Noise buildup usually introduced in the recording process is dramatically reduced; however, as with all noise-reduction systems, any noise present in the original audio signal is not reduced.

FIGURE 9.4 dbx® noise-reduction system. (Courtesy of dbx®, a division of AKG Acoustics, Inc.)

9.10 REVERBERATION

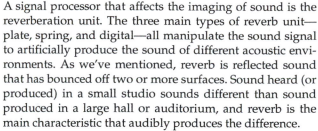

A signal processor that affects the imaging of sound is the reverberation unit. The three main types of reverb unit—plate, spring, and digital—all manipulate the sound signal to artificially produce the sound of different acoustic environments. As we've mentioned, reverb is reflected sound that has bounced off two or more surfaces. Sound heard (or produced) in a small studio sounds different than sound produced in a large hall or auditorium, and reverb is the main characteristic that audibly produces the difference.

A **plate reverb** unit consists of a metal sheet suspended within a frame. A transducer at one end of the plate changes the audio signal into mechanical energy and vibrates the plate. A contact mic at the other end of the plate picks up the vibrations (reverb), then sends the altered signal wherever it needs to go. Plate reverb systems are rather bulky and expensive and aren't found in many production facilities.

Spring reverb units are less expensive and operate in a similar manner. A transducer at one end of a coiled wire spring causes the spring to vibrate when a signal is sent through it. A contact mic at the other end picks up the altered signal from the spring. The amount of reverb introduced into the signal can usually be controlled on both plate and spring reverb units.

The modern production studio is more likely to have a **digital reverb** unit. Rather than being a mechanical-electrical device, the digital reverb is a pure electronic device (see Figure 9.5). The original signal is fed into the unit and is electronically processed to achieve the reverb effect; then the altered signal is sent out of the unit. Usually a greater number of effects can be produced with the digital units.

9.11 DIGITAL DELAY

A **digital delay** unit can be used in both the production studio and on-air control room. As its name implies, this signal processor actually takes the audio signal, holds it, and then releases it to allow the signal to be used further. The time the signal is held or delayed can be varied from fractions of a second to several seconds. While there are analog delay units, most systems are digital, and the incoming signal is converted from its analog form to digital for processing and back to analog after processing.

In the on-air studio, a delay unit is often used in conjunction with a telephone talk show. The program signal is sent through the delay unit to provide approximately a 7-second delay before it is sent to the transmitter. If something is said by a caller that should not be broadcast, the operator has time to "kill" the offending utterance before it is actually broadcast. (This is why callers are asked to turn their home radio down if they're talking to an announcer on-air because the sound they hear on the radio is the delayed sound and not the words they are actually saying into the tele-

FIGURE 9.5 Digital reverb unit. (Courtesy of Yamaha Professional Audio Division)

phone. It's extremely difficult to carry on a conversation when you can hear both sounds.)

In the production studio, delay units are used to create special effects similar to reverb. Set for an extremely short delay, the units can create an effect that sounds like a doubled voice or even a chorus of voices.

9.12 DYNAMIC RANGE

When we mention **dynamic range** in a radio production context, we're referring to the range of volumes of sound that broadcast equipment can handle. This intensity of a sound is measured in decibels (dB). One decibel represents the minimum difference in volume that we can hear, but a change of 3 decibels is often necessary before we actually perceive a difference in volume.

The dynamic range goes from 0 decibels at the **threshold of hearing** to 120+ decibels at the **threshold of pain**. A whisper is around 20 decibels, normal conversation is near 60 decibels, and shouting is about 80 decibels. Average music-listening levels are between 30 and 80 decibels, but some rock concerts have been measured above 110 decibels.

Dynamic range also relates the volume of one signal to another, such as signal to noise. We should note that dynamic range is measured on a logarithmic scale. For example, to hear one audio signal twice as loud as another, you would have to increase the volume of one by 10 decibels in relation to the other.

A 60-decibel dynamic range was once considered quite adequate for high-quality broadcast production equipment; today's digital equipment, however, offers an increased dynamic range and a 90-decibel range is now common.

9.13 COMPRESSORS

Two signal-processing devices, the compressor and the limiter, are used to affect the dynamic range of the audio signal. Although they are most often used to process the

signal between the studio and the transmitter and therefore aren't pure production room devices, they were the first processing devices used in radio, and virtually every radio station uses them. They are also occasionally used in the production room to process the signal before it's sent to an audio tape recorder.

The **compressor** operates as an automatic volume control. If the audio signal is too loud, the compressor automatically lowers it, and if the signal is too soft, the compressor increases it. Several adjustments on the compressor determine its actual operation. The threshold of compression is the setting of the level of signal needed to turn on the compressor. As long as the audio signal stays below this point, the compressor doesn't do anything. Input level can be adjusted on the compressor, although this level is also determined by the output of the audio signal source.

In any case, the compressor really needs to work only some of the time. If the input is too low, the threshold of compression will never be reached, and if the input is too high, the compressor will severely restrict the dynamic range. The compression ratio determines how hard the compressor works. A ratio of five-to-one means that if the level of the incoming signal increases ten times its current level, the output of that signal from the compressor will only double. Compressors also have settings for attack time (how quickly volume is reduced once it exceeds the threshold) and release time (how quickly a compressed signal is allowed to return to its original volume). The release time adjustment is very important because too fast a setting can create an audible pumping sound as the compressor releases, especially if there is a loud sound immediately followed by a soft sound or a period of silence.

9.14 LIMITERS

A **limiter** is a form of compressor with a large compression ratio of ten-to-one or more. Once a threshold level is reached, a limiter doesn't allow the signal to increase anymore. Regardless of how high the input signal becomes, the

FIGURE 9.6 Compressor/limiter. (Courtesy of Symetrix, Inc.)

FIGURE 9.8 Equalizer screen from DAW. (Courtesy of Innovatice Quality Software)

output remains at its preset level. If they're adjustable, attack and release times on limiters should be quite short.

Both the limiter and the compressor can be rather complicated to adjust properly. Too much compression of the dynamic range makes an audio signal that can be tiresome to listen to, and pauses or quiet passages in the audio signal are subject to the pumping problem mentioned earlier. Since they're often associated with the transmitter only, compressors and limiters are usually the domain of the engineer. If you do have access to them in the production facility, it may take some experimenting to get the kind of processing that you're looking for because there are no standard settings for signal-processing devices. Figure 9.6 shows a system that combines both a compressor and limiter in a single unit.

9.15 MULTI-EFFECTS PROCESSORS

Currently there's a general trend to build signal-processing devices so that they perform more than one function—that is, to put more digital effects in one black box. For example, Eventide's DSP-4000B **Ultra-Harmonizer**, Yamaha's SPX-990, and Lexicon's PCM-80 are all popular signal processing tools in the production studio (see Figure 9.7). These devices offer a variety of audio effects in one unit, including the ability to alter the pitch of an incoming audio signal, time compression and expansion, delay, natural reverb effects, flanging, time reversal, and repeat capabilities. Any production person should find it enjoyable to experiment with the variety of creative effects that can be produced with a multi-effects processor.

Another form of multi-effects processor is the digital audio workstation that we mentioned in the chapter on audio tape editing. In addition to offering improved editing

capabilities, most workstations have provisions for adding special effects to the audio signal. When you're in the multitrack screen on the SAW Plus system, you can add an effect to the audio on any track by mouse clicking on an "EFX" button. This opens a window that allows you to select the effect you want. This system offers effects like simple reverb, phasing, equalization (see Figure 9.8), and even reverse audio. You can add one effect or several to the same audio signal by just "patching" in the effects you want. Other workstation systems operate in a similar manner and in many cases are simpler to set up and operate than their black box counterparts.

9.16 OTHER SIGNAL PROCESSORS

A **flanger** is another processor for producing a specific special effect. This unit electronically combines an original signal with a slightly delayed signal in such a way as to cause an out-of-phase frequency response that creates a filtered swishing sound. A **de-esser** is an electronic processor designed to control the sibilant sounds without affecting other parts of the sound signal. As the name implies, a **stereo synthesizer** is a processor that takes a mono audio signal (input) and simulates a stereo signal (output). Some processors use a form of delay to provide separation between the left and right channel outputs; others use a form of filtering to send certain frequencies to one channel and certain other frequencies to the other channel to provide a synthetic stereo effect.

9.17 CONCLUSIONS

This chapter is not a complete guide to signal-processing equipment. There are other units available and in use in radio production facilities. Nor is this chapter intended to make you a professional operator of such equipment. What is intended is that you become aware of a number of the more common processors and that you have an understanding of their basic purpose. The actual operation of most of this equipment will take some trial-and-error work in your production facility. Above all, remember that one

FIGURE 9.7 Multi-effects processor. (Courtesy of Eventide, Inc.)

small effect used as part of a production can be very effective, but too many special effects in a production take away from their specialness. Use signal processing in moderation, and bear in mind that a lot of great radio production has been produced using no signal-processing equipment at all.

Self-Study

■ QUESTIONS

1. The equalizer processes an audio signal by altering which of the following?
   a) volume
   b) imaging
   c) dynamic range
   d) frequency response

2. An audio signal that has been equalized would be called a "dry" signal.
   a) true
   b) false

3. Which type of equalizer can select an exact center frequency and bandwidth as well as alter the volume at that frequency and bandwidth?
   a) graphic
   b) parametric
   c) dielectric
   d) full-octave, ten-band

4. What type of filter would most likely be used to attenuate or eliminate a 60-hertz hum in a recording?
   a) low pass filter
   b) band pass filter
   c) notch filter
   d) low cut filter

5. The 60-hertz hum mentioned in question 4 could also have been eliminated by the use of either Dolby or dbx noise reduction.
   a) true
   b) false

6. A signal processor that affects the imaging of a sound is the _____.
   a) equalizer
   b) noise-reduction unit
   c) de-esser
   d) reverb unit

7. In order to create a tinny voice, you would _____.
   a) cut out most of the lower frequencies
   b) cut out most of the higher frequencies
   c) eliminate the EQ
   d) increase compression

8. The noise-reduction system most likely to be found in the radio production studio is _____.
   a) Dolby A
   b) Dolby S
   c) Type I
   d) Eventide Harmonizer

9. In the Dolby system of noise reduction, _____.
   a) volumes of certain frequencies are increased during recording and decreased during playback
   b) the dbx is increased with a calibrated tone so that it attains the level of 30 decibels
   c) all frequencies pass through, except ones that have been preset by the notch filter
   d) once a threshold level is reached, the signal isn't allowed to increase anymore

10. Which type of reverb has only electronic, not mechanical, elements?
    a) plate
    b) spring
    c) digital
    d) unity-gain

11. A compressor _____.
    a) usually has a compression ratio of ten-to-one
    b) lowers a signal that's too loud and raises one that's too soft
    c) doesn't operate unless it's connected to a digital delay unit
    d) produces an out-of-phase, filtered, swishing sound

12. Any signal-processing equipment that's labeled a unity-gain device would amplify all frequencies of the signal going through that equipment an equal amount.
    a) true
    b) false

13. Which of the following is *not* a type of Dolby noise reduction?
    a) Dolby S
    b) Dolby B
    c) Dolby C
    d) Dolby D

14. Which signal-processing device is most likely to offer a variety of effects, such as pitch change, time compression, reverb, and flanging?
    a) dbx Graphic Equalizer
    b) Eventide Broadcast Delay
    c) dbx Compressor/Limiter
    d) Eventide Ultra-Harmonizer

15. Which signal-processing device inputs a mono signal and outputs a simulated stereo signal?
    a) digital delay
    b) band cut filter
    c) flanger
    d) stereo synthesizer

16. Today's digital equipment offers an increased dynamic range and a _____ decibel range is now common.
    a) 120
    b) 90
    c) 60
    d) 3

17. A digital audio workstation usually includes digital reverb and delay effects, but rarely any other type of signal processing capability.
    a) true
    b) false

18. Which graphic equalizer setting would most likely be used to add "brilliance" to an announcer's voice?
    a) boost at 60 Hz
    b) cut at 250 Hz
    c) boost at 1 kHz
    d) cut at 8 kHz

19. Which signal processor is designed to help control sibilance in an announcer's voice?
   a) compressor
   b) limiter
   c) de-esser
   d) spring reverb

20. As a final test on signal processing, match the following pieces of equipment (1, 2, 3 . . .) with their primary functions (f, d, i . . .). (*Letters may be used more than once.*) Then select the correct set of answers from the sequences in a, b, c, or d below.

   1. _____ parametric equalizer
   2. _____ plate reverb unit
   3. _____ limiter
   4. _____ Dolby
   5. _____ digital reverb unit
   6. _____ dbx
   7. _____ compressor
   8. _____ notch filter

   f. manipulates the frequency range
   d. changes dynamic range
   i. deals with imaging
   n. reduces noise

   a) 1.d 2.i 3.f 4.n 5.i 6.n 7.f 8.d
   b) 1.f 2.i 3.d 4.n 5.i 6.n 7.d 8.f
   c) 1.d 2.n 3.n 4.f 5.d 6.i 7.n 8.f
   d) 1.f 2.n 3.d 4.i 5.n 6.i 7.d 8.f

■ ANSWERS

If You Answered A:
   1a. No. You're headed in the right direction. Equalizers do increase or attenuate volumes at specific frequencies, but they do this to alter another sound characteristic. (Reread 9.1 and 9.2.)
   2a. No. A dry audio signal is an unprocessed signal. (Reread 9.1.)
   3a. No. Center frequencies and bandwidths are preset on graphic equalizers. (Reread 9.3 and 9.4.)
   4a. No. This type of filter allows lower frequencies to pass and would not eliminate noise at 60 hertz. (Reread 9.6.)
   5a. No. Noise-reduction units, regardless of brand name, can't eliminate noise that already exists in a recording. They only prevent noise during the recording process. (Reread 9.7.)
   6a. No. Equalizers affect frequency response. (Reread 9.2 and 9.10.)
   7a. Correct. Cutting the bass will give a tinny sound.
   8a. No. (Reread 9.8 and 9.9.)
   9a. Correct. It's a two-step process.
   10a. No. It has a transducer that changes the audio signal into mechanical energy. (Reread 9.10.)
   11a. Wrong. That's a limiter. (Reread 9.13 and 9.14.)
   12a. No. A unity-gain device doesn't amplify the overall level of the incoming signal at all. (Reread 9.2.)
   13a. Wrong. This is one of the newest and most sophisticated Dolby noise-reduction systems. (Reread 9.8.)
   14a. No. This doesn't really exist. It's a mixture of signal processing. (Reread 9.3, 9.9, and 9.15.)
   15a. No. You may be thinking about a stereo synthesizer that uses a form of delay. (Reread 9.11 and 9.16.)
   16a. No. You may be getting this confused with the threshold of pain. (Reread 9.12.)
   17a. No. Most DAW's have multi-effect capability. (Reread 9.1 and 9.15.)
   18a. No. This would punch up the bass. (Reread 9.3.)
   19a. Wrong. This is a form of automatic volume control used to affect dynamic range. (Reread 9.13 and 9.16.)
   20a. No. You're confusing frequency response and dynamic range. (Reread 9.2–9.5 and 9.12–9.14.)

If You Answered B:

1b. No. Imaging can be affected by other signal processors. (Reread 9.1 and 9.2.)

2b. Correct. This is the right response.

3b. Yes. The parametric equalizer gives the operator the greatest control over the EQ process.

4b. No. This type of filter is usually used to allow a range of frequencies to pass, not to eliminate a single frequency. (Reread 9.6.)

5b. Yes. Because you can't eliminate existing noise with noise-reduction units.

6b. No. Noise-reduction units affect dynamic range. (Reread 9.7 and 9.10.)

7b. No. That is where the tinny sound would be. (Reread 9.5.)

8b. Correct. It's one of the newer common ones.

9b. No. (Reread 9.8 and 9.9.)

10b. No. It has a transducer that changes the audio signal into mechanical energy. (Reread 9.10.)

11b. Yes. It lowers and raises signals in that manner.

12b. Yes. Unity-gain devices don't amplify the overall level of the incoming signal.

13b. Wrong. Dolby B is a common noise-reduction system found on cassette recorders. (Reread 9.8.)

14b. No. This isn't correct. (Reread 9.11 and 9.15.)

15b. No. You may be thinking about a stereo synthesizer, which uses a form of filtering. (Reread 9.6 and 9.16.)

16b. Correct. This is the right response.

17b. Yes. Most DAW's have multi-effect capability.

18b. No. This would minimize bass boominess. (Reread 9.3.)

19b. Wrong. This is a form of automatic volume control used to affect dynamic range. (Reread 9.14 and 9.16.)

20b. Correct. You have now finished the section on signal processing.

If You Answered C:

1c. No. Dynamic range can be affected by other signal processors. (Reread 9.1 and 9.2.)

3c. There's no such thing. (Reread 9.3 and 9.4.)

4c. Correct. This type of filter allows all frequencies to pass except a specified one, which we could specify at 60 hertz to eliminate the hum.

6c. Wrong. A de-esser is an electronic processor designed to control sibilant sounds. (Reread 9.10 and 9.16.)

7c. Wrong. That wouldn't really be possible. (Reread 9.5.)

8c. No. (Reread 9.8 and 9.9.)

9c. No. You are confusing this with filters. (Reread 9.6 and 9.8.)

10c. Correct. It has no mechanical elements.

11c. No. They have nothing to do with each other. (Reread 9.11 and 9.13.)

13c. Wrong. Dolby C is a common noise-reduction system found on cassette recorders. (Reread 9.8.)

14c. No. This isn't correct. (Reread 9.9 and 9.13–9.15.)

15c. No. You're thinking about another signal-processing device. (Reread 9.16.)

16c. Wrong. You may be getting this confused with the level of normal conversation. (Reread 9.12.)

18c. Yes. This would add brilliance to a voice.

19c. Right. This is the correct answer.

20c. You're either lost or you're guessing. (Reread the entire chapter.)

If You Answered D:

1d. Correct. Equalizers allow you to adjust selected frequency volumes and thus alter the audio signal's frequency response.

3d. No. This would be a type of graphic equalizer. (Reread 9.3 and 9.4.)

4d. While it could eliminate a 60 hertz hum, low cut filters normally are designed to eliminate *all* frequencies below a certain point. There's a better response. (Reread 9.6.)

6d. Yes. This is what reverb units do by electronically changing the apparent acoustic environment in which we hear the sound.

7d. No. A compressor acts as an automatic volume control affecting the dynamic range of an audio signal. (Reread 9.5 and 9.13.)

8d. No. This is a multi-effect device, not a noise-reduction unit. (Reread 9.7–9.9 and 9.15.)

9d. Wrong. This happens in a limiter, but not during noise reduction. (Reread 9.8 and 9.14.)

10d. No. Unity-gain relates to equalizers, not reverb. (Reread 9.2 and 9.10.)

11d. No. You're thinking of a "flange" effect. (Reread 9.13 and 9.16.)

13d. Right. This is not a current Dolby noise-reduction system.

14d. Yes. This is a multi-effect processor.

15d. Correct. As the name implies, a stereo synthesizer makes a mono signal a "fake" stereo signal.

16d. Wrong. You're thinking about the amount of change in volume that's usually necessary before we actually hear a difference in level. (Reread 9.12.)

18d. No. This would decrease the highs, if anything. (Reread 9.3.)

19d. Wrong. This is a type of processor used to affect imaging of an audio signal. (Reread 9.10 and 9.16.)

20d. Wrong. You're confusing imaging and noise reduction. (Reread 9.7–9.11.)

Projects

■ PROJECT 1

Operate the signal-processing equipment in your facility.

Purpose

To familiarize you with the various functions of signal-processing equipment.

Advice, Cautions, and Background

1. You may not be able to do this entire project because of lack of equipment. Do it as best you can. If you don't have access to any signal-processing equipment, try to use equalizers that may be part of your audio console.
2. You can prepare the tape you'll need for this at home if you have a mic, a turntable or CD, and a tape recorder.
3. Turn in both the original tape and the completed project tape, so your instructor can compare them.
4. Feel free to review material in this book that deals with CDs, audio consoles, and tape recorders so that you can do this project properly.
5. You may need help from your instructor or engineer to arrange the signal-processing equipment.

How to Do the Project

1. Make a tape that consists of you talking for at least a minute, introducing a song, then the song itself. While you're recording the music, raise the volume at one point so that it's overmodulated, and then lower the volume at another point so that it's riding in the mud.
2. Rerecord this tape, using whatever signal processing you have available to accomplish the following:
   a. Change your voice for about 15 seconds so that it sounds tinny.
   b. Accentuate the bass notes of the music for a brief period of time, and then eliminate or lessen the bass notes.
   c. Lower the signal that's too loud and raise the signal that's too soft.
3. Clearly label the original tape and the rerecorded tape with your name and SIGNAL PROCESSING. Turn in both to your instructor for credit for this project.

■ PROJECT 2

Visit a radio station or recording studio, and learn about the signal-processing equipment used there. Write a report about what you learn.

Purpose

To give you more familiarity with signal-processing equipment.

Advice, Cautions, and Background

1. Find a place that has enough signal-processing equipment to make your trip worthwhile. Your instructor may arrange a trip for the entire class.

2. Find someone at the facility who's willing to spend some time with you explaining how the equipment works.
3. Prepare the questions that you want to ask so that you can guide the visit to some extent.
4. If possible, experiment with some of the equipment yourself. If the facility is unionized, you won't be able to do this. Don't push on this particular point because some people are very sensitive about who touches their equipment, but don't back away from any opportunity.
5. Once you have an appointment, keep it, and arrive on time.

How to Do the Project

1. Call various radio stations and recording studios until you find one that's both willing to let you visit and has sufficient signal-processing equipment.
2. Visit the facility, and talk with someone who can give you information for your report. Make a list of all the equipment you are shown, and take notes regarding it. Some of the things you may want to find out are:
   a. Is the equalizer graphic or parametric?
   b. Do they have band pass or notch filters or both?
   c. Do they use Dolby or dbx and, if so, which system?
   d. Is the reverb system plate, spring, or digital?
   e. Is there a digital delay unit, and, if so, how many seconds' delay does the station use?
   f. Do they have a compressor? A limiter?
   g. Do they have any multi-effects signal processing equipment?
   h. What other signal-processing equipment do they have, and what does it do?
3. As soon as you leave the facility, organize your notes so that you remember the main points.
4. Write a report. It should be several pages long, preferably typed. Write your name and SIGNAL-PROCESSING EQUIPMENT TOUR on a title page.
5. Turn in the report to your instructor for credit for this project.

■ PROJECT 3

Record and play back material, using Dolby noise reduction.

Purpose

To make you aware of some of the characteristics of noise-reduction signal processing.

Advice, Cautions, and Background

1. You'll need a recorder with Dolby noise-reduction capability to complete this project.
2. You'll also need a prerecorded tape that has been encoded with Dolby.
3. Other configurations can be used, but this project will assume that a cassette recorder with Dolby B is being used.

How to Do the Project

1. Set up the cassette recorder to play back the prerecorded cassette tape with Dolby B. Make sure the packaging of the cassette indicates that it was recorded with Dolby.
2. Listen to the tape for several minutes. Stop the tape and rewind it to the start point.
3. Turn the Dolby B switch to the off position, and play the tape again, listening for a comparable amount of time.
4. Note any differences in the sound quality of the recording as it's played back with and then without Dolby.
5. On a blank cassette tape, record several minutes of an announcer talking. Make sure the Dolby B switch is in the on position.
6. Listen to this recording. Then make a similar recording with the Dolby B switch in the off position.
7. Listen to this recording, and note any differences in the sound quality between the recording with and without Dolby noise reduction.
8. If your recorder also has Dolby C, repeat steps 5–7 with the Dolby C switch on.
9. Write a short paper that describes any differences you noted. Label the paper DOLBY DIFFERENCES.
10. Turn in this paper and the tape you made to your instructor to receive credit for this project.

CHAPTER TEN

The Digital Production Studio

Information

10.1 INTRODUCTION

Digital technology has revolutionized how the radio production person can record, edit, and otherwise manipulate an audio sound signal. Digital technology has also ushered in tremendous improvements in sound quality—better frequency response range, improved signal-to-noise ratio, increased dynamic range, and lower noise and distortion, to name a few. (If these sound-quality terms seem unfamiliar to you, check the glossary; and if you've forgotten about the differences between digital and analog recording, review section 4.2 before continuing with this chapter.)

From the advent in the early 1980s of the compact disc player, to the development in the late 1980s of the digital audio workstation, to the mid-1990s total digital production studio and radio station, radio has eagerly embraced digital technology. This chapter introduces some additional digital audio equipment used in the radio production studio and some thoughts about what the future may hold in this area.

10.2 THE DIGITAL "CART MACHINES"

Designed to replace standard cart machines, **digital cartridge recorders**, which look and operate like analog cart recorders, have been developed by several broadcast manufacturers. Digital cart machines are priced in the same range as analog units and come in both record/play and play-only configurations. Instead of an analog audio cart as the recording or storage medium, one of these machines (see Figure 10.1) uses a standard 3.5" computer disk. Readily available in most office supply stores, these disks come in 2 MB and 13 MB capacity, which can provide about 5 minutes of high-quality stereo audio.

To increase recording time or add features, some machines (see Figure 10.2) use a specialized high-density magneto-optical disk cartridge known as a Bernoulli disk in place of the standard computer disk.

Some machines employ an internal hard-disk drive (see Figure 10.3) that allows instant access to up to 1000 "cuts" stored on the drive. A "find" feature allows the operator to quickly locate a cut just by typing in the name of the track.

Or you can scroll through the entire contents of the drive. Hard drives offer the greatest amount of storage space, extremely quick access time, and low maintenance. However, if a hard drive device "crashes," a huge amount of audio can be lost, which can disrupt a production or on-air studio.

Most digital cart recorders offer selectable sampling rates and can add an electronic "label" (through the use of a keyboard), which can show the cart "name" and "timing" information on an LCD (liquid crystal display) screen.

10.3 THE MINIDISC

The form of digital cart machine that has perhaps found the strongest presence in the radio production studio is the **MiniDisc (MD)**. Originally developed by Sony, the MiniDisc can be a handheld, tabletop, or rack-mount system (see Figure 10.4) that comes in a record/play or play-only configuration. Employing a tiny 2.5-inch computer-type disk, the MD can still hold up to 74 minutes of music because of a data compression scheme. The MiniDisc is *not* actually CD quality, but extremely high-quality audio. The MD also features a "shock absorber" system that uses a memory buffer to store music that can continue to play for a few seconds if the player mistracks until the pick-up can return to its correct position.

The portability of some MD models and its "near digital" sound makes it a serious competitor for standard cassettes and DAT recorders in some production situations. The latest development in the MiniDisc area is a small, single component (see Figure 10.5) that combines an MD recorder with a small mixing console to provide recording, editing, and mixing features. Over 30 minutes of digital recording time on four tracks is available, and since it is digital, the audio can be moved from track to track with no noise buildup or quality loss.

10.4 DATA COMPRESSION

In order for the MiniDisc to record the tremendous amount of necessary digital data and still employ a small computer-type disk recording medium, the MD recording process

FIGURE 10.1   Digital cartridge recorder. (Courtesy of Fidelipac Corporation)

uses a data **compression** system developed by Sony known as **ATRAC (Adaptive Transform Acoustic Coding)** to provide digital quality sound. Like any other data compression scheme, data compression occurs because only audible sounds are encoded. In other words, sound below the threshold of hearing or too soft to hear and sounds masked by louder sounds heard at the same time are excluded, according to a "psycho-acoustic model." In addition, those sounds that are encoded use only the number of data bits necessary for near-CD quality sound. During recording, the incoming audio signal is first divided into over 50 narrow frequency subbands. Next, each of the subbands is ana-

lyzed and encoded, according to the ATRAC scheme. For each digital sample, the subband is allocated a certain number of data bits; subbands containing complex signals are encoded with many data bits, and other subbands containing mostly inaudible or masked sound are encoded with only a few data bits. Data compression occurs because the amount of data that must be recorded onto the MD is greatly reduced if all the inaudible sounds are simply ignored and not encoded. This allows the MiniDisc recording to cut down the amount of data that must be recorded by about one fifth with no apparent sound quality loss.

There is no standard method of data compression, and

FIGURE 10.2   Magneto-optical disk recorder. (Courtesy of Studer)

FIGURE 10.3   Hard disk system. (Courtesy of 360 Systems)

other digital equipment utilizes other types of audio data reduction schemes. There has been some concern of the audible effects on audio that has been compressed more than once. For example, a MiniDisc could be played through a signal processor that also uses data compression. Some tests have shown a degradation of the audio signal with some combinations of data compression; however, this doesn't appear to be a major problem.

10.5   THE DIGITAL AUDIO EDITOR

The least expensive **digital audio editor** utilizes a standard IBM-compatible, Macintosh, or Amiga computer, plus some specialized equipment, to provide an audio editing system for around $1000 (assuming the station already has the computer). Often referred to as "desktop radio," this type of production system has the potential of replacing an entire existing production studio. While some systems may

run on a very basic computer setup, most will require a Pentium 90 or higher speed CPU, 16 MB of RAM (random access memory), a mouse driver, a high-resolution VGA monitor, and a fast hard-disk drive with about 10 megabytes of memory for each stereo minute of audio. The specialized equipment includes a **digital signal processor (DSP) audio card** (see Figure 10.6), a computer software program to perform the actual editing, and some type of interface or I/O (input/output) device to move the audio signal from its source in the production studio into the editing system, then back into the production studio.

A simple editing computer program merely defines segments of audio and then assembles the segments into a list to be played back. For example, the background to a "concert spot" might include several songs by the featured group. Segments of each song are recorded on the editor's hard disk as a **soundfile**. The exact portion of each segment that is to be used in the spot can be selected, using the mouse to define an **event** in a fashion similar to standard

FIGURE 10.4   MiniDisc recorder. (Courtesy of Denon Electronics)

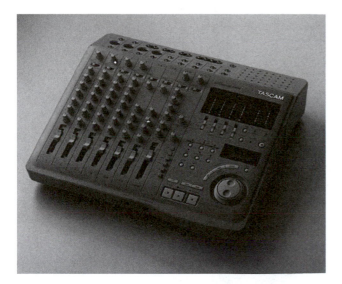

FIGURE 10.5 MD multitrack recorder. (Courtesy of TEAC America, Inc.)

editing, i.e., selecting a start point and end point for each "edit." In this case, the edit marks are made on the computer screen, which shows a "picture" of the audio (usually the actual waveform) and can be easily moved, or "trimmed," to the exact position desired (see Figure 10.7). The final step is to assemble the events into a **playlist**, or order in which you want them played. The various edits can be played back one event at a time or in a seamless montage.

One of the biggest advantages of this type of system is the *nondestructive* nature of the editing. The original audio isn't actually altered; edit information (where to start and stop playing the audio) is recorded in an "edit decision list." Some systems do editing that's *destructive* because portions of the audio are actually deleted or cut out; however, even these systems usually allow you to preview the edit before it's performed and undo it if isn't marked properly.

More sophisticated editing programs allow the recording of several "tracks" (just like a multitrack recorder) that can then be manipulated individually and ultimately mixed together to provide a completely produced spot. The next chapter on multitrack recording will examine these techniques.

Another type of digital audio editor that *doesn't* need a computer, mouse, monitor, or special plug-in card is beginning to see a lot of popularity in the production studio. 360 Systems Short/cut™ personal audio editor is configured in a compact, portable case as shown in Figure 10.8 and records audio directly to its internal hard disk. Beneath a two-track waveform display, are three distinct sections of controls. To the right, are the basic transport controls (like play, pause, record, etc.), in the center are editing controls (like edit in, cut, erase, and a large scrub wheel), and to the left, are the system controls (like hot keys, menu, directory, and a QWERTY keyboard for labeling functions). Editing with the Short/cut™ is as easy as recording or loading in a file, scrubbing to the edit in point, and then scrubbing to the edit out point. The marked section of audio will become highlighted and can then be removed, copied, erased, or new material can be inserted. Like most digital equipment,

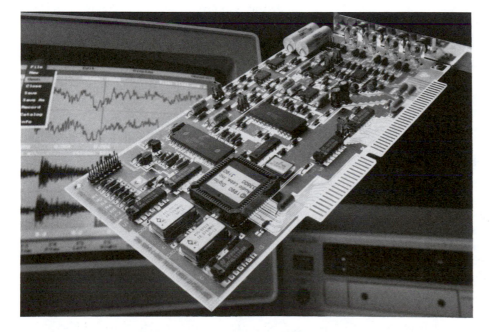

FIGURE 10.6 DSP audio card. (Courtesy of Digital Audio Labs)

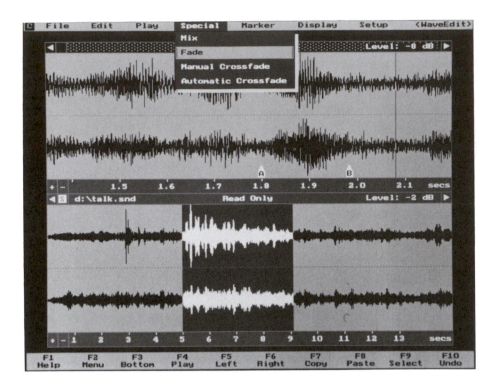

FIGURE 10.7   Screen from audio editing system. (Courtesy of Digital Audio Labs)

if you make a mistake an undo function will allow you to correct it. The Short/cut™ has a selectable 44.1 kHz or 48 kHz sample rate and 360 Systems offers 1½ hour or 3 hour recording time models. Less complex than some of the PC-based audio editors, this type of digital audio editor is being touted as a digital replacement for the reel-to-reel recorder and may become very commonplace in the digital production studio of the future.

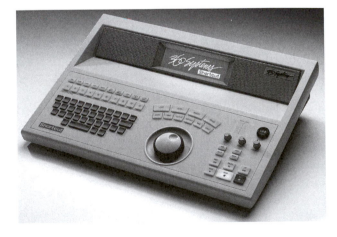

FIGURE 10.8   Personal audio editor. (Courtesy of 360 Systems)

10.6 DIGITAL AUDIO WORKSTATIONS

A further extension of digital editing technology in the radio production studio has been the development of the **DAW**, or **digital audio workstation** (see Figure 10.9). These are hard-disk-based systems that incorporate computer keyboards, touch screens, or mice to manipulate music, dialogue, and sound effects. The original sound signal is converted to digital form and can then be stored, manipulated, and recalled using the workstation.

The basic components of the DAW are similar to the elements that make up the simple digital audio editor mentioned in section 10.5; however, the computer is usually faster and more powerful than the standard personal computer, and DAWs are often "packaged" as a single unit similar to that shown in Figure 10.10. A **frame** houses the computer chassis, power supply, and motherboard. A large-capacity **hard disk drive** is required to store and manipulate the digital audio data. Remember, each stereo minute of audio will "eat up" about 10 megabytes; however, data compression techniques will no doubt reduce the capacity requirements of the next generation of workstations.

A **user interface**, such as a keyboard, touchscreen, or mouse, is necessary to actually operate the workstation. However, many workstations have tried to be more user-friendly to radio production people by making the interface include typical radio production elements, such as faders, cue wheels, and tape-transport buttons like "rewind" or

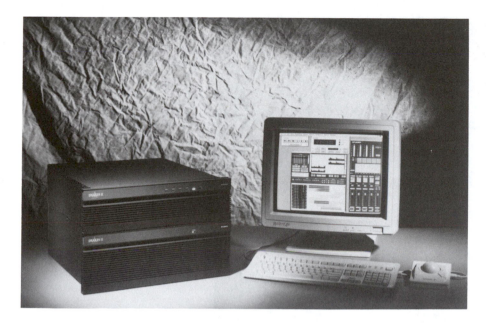

FIGURE 10.9 Digital audio workstation. (Courtesy of Studer Revox America, Inc.)

"play." Often these interfaces are shown in graphics form on the computer screen, but some systems actually attach to an audio board (refer to Figure 10.10 again). An **audio card** provides the connection between the workstation and the other audio equipment by converting analog audio into digital data and vice versa. Through the use of serial and parallel ports, most workstations have the ability to network with other workstations or interface with printers, modems, and other components.

10.7 MIDI/SMPTE TIME CODE IN RADIO PRODUCTION

Most digital audio workstations and many other pieces of digital broadcast equipment have the ability to incorporate MIDI and SMPTE synchronization. **MIDI** (**musical instrument digital interface**) is an interface system that allows electronic equipment—mainly musical instruments like synthesizers, drum-machines, and samplers—to "talk" to each other through an "electronic" language.

SMPTE (Society for Motion Picture and Television Engineers) **time code** is an electronic language developed for video tape editing that identifies each video "frame" with an individual address. The time code numbers consist of hour, minute, second, and frame. The frame digits correspond to the 30 video frames in each second.

Both MIDI and SMPTE signals can be used to reference various individual pieces of equipment and accurately start, combine, or stop them. Someone working in the production studio is more likely to make use of MIDI or SMPTE than someone in the on-air broadcast studio. MIDI,

in particular, is often used to interface various musical sounds for a commercial or public-service announcement.

10.8 ADVANTAGES OF THE DIGITAL PRODUCTION STUDIO

One of the main advantages of digital equipment like the CD player, DAT recorder, and DAW is improved audio signal quality. The digital format offers superior technical specifications in the area of frequency response, dynamic range, signal-to-noise ratio, wow, flutter, and other forms of audio distortion. Not only does this provide an improved initial audio signal, but the digital process doesn't build up any added noise during recording, dubbing, or transmitting.

Production people quickly realized another advantage of digital editors like the DAW was the ability to edit without using razor blades to physically cut the audio tape. A nondestructive editing method provides a safer work environment in the production studio, plus it allows you to preview the editing. Most DAWs have an UNDO button to put the sound back into its original form if you don't like the way the edit came out. Another operational advantage of the DAW is a fast random access, or the ability to immediately cue up at any point of all the material stored in the system. Faster and easier operation of digital equipment should lead to less time spent on basic production functions and more time spent on the creative aspect of the production process.

It is also easier to undertake many editing operations with digital editing than with traditional razor blade edit-

FIGURE 10.10 "User-friendly" DAW (broadcast style). (Courtesy of AKG Acoustics, Inc.)

ing. For example, if you must remove a very short segment, such as someone momentarily stuttering on a word, it's easy to do because you can see the waveform, make it larger, and then remove the stuttering section accurately. Also, if you record something and then need to make it slightly shorter or longer, you can do so easily. For instance, if you record what is to be a 15-second commercial and it turns out to be 17-seconds long, you can remove short bits of silence throughout the commercial to get it to 15 seconds. In fact, there are computer programs that do this for you. You tell it how long you want something to be, and the program adjusts the material accordingly.

Digital equipment, such as the CD and digital cart recorders, offers a labeling function that's more convenient and comprehensive than paper labels. IDs and other types of label information can be encoded within the digital media. A "table of contents" can provide names, times, and other data through a front-panel window on the equipment or through the video monitor screen of a DAW.

Another technical advantage of digital is fewer maintenance problems, such as dirty tape heads or alignment problems. Most digital equipment promises a longer interval between breakdowns than comparable analog equipment, and when there's a technical problem, digital equipment usually takes less time to repair. Since there are fewer internal operating parts in digital equipment, in many cases repair is a simple replacement of one circuit board or component for another.

10.9 DISADVANTAGES OF THE DIGITAL PRODUCTION STUDIO

Most of the disadvantages of digital equipment are tied more to their newness than to inherent problems in the systems. For example, learning to operate a DAW can take several months, a factor that makes them somewhat user unfriendly. Regardless of what manufacturers might claim, they do take some time to learn, and production people must be willing to give that time. Stations might find that some staff resist this type of change of procedures in radio production techniques.

It would be hard to argue against the superior sound quality that digital equipment provides because it can be readily measured and compared to analog sound. Nonetheless, a criticism of digital sound still exists because, for some, the clear, brilliant sound has been likened to a "cold" or "mechanical" sound. It's not what we're hearing, but rather what we *don't* hear that our ears are used to with old analog equipment. The tremendous frequency response, dynamic range, and lack of noise and other forms of distortion, for some people, is too "harsh." Of course, this judg-ment of sound quality is highly subjective, and for many, if not most, people, digital sound is just fine.

The high cost of digital equipment originally was considered a drawback. Like most new technologies, however, the cost factor has decreased with continued development. While CD players can be purchased in the $100 range, basic DAWs will still cost a station more than $10,000, and full-blown DAW-based digital radio stations will cost much more. The systems that can be put together with PC-based consumer-grade equipment offer the possibility of an "entry-level" approach to getting started with digital production, but not all systems can be used successfully for extensive audio projects.

Some production facilities will have to look carefully at the space requirements of new digital equipment. While some digital equipment is smaller than the analog equipment it's replacing, additional counter space for a video monitor and mouse will be necessary, as will additional rack space for hard-disk drives and other digital equipment. For a time, some new digital hardware will have to coexist with the traditional broadcast equipment in the studio.

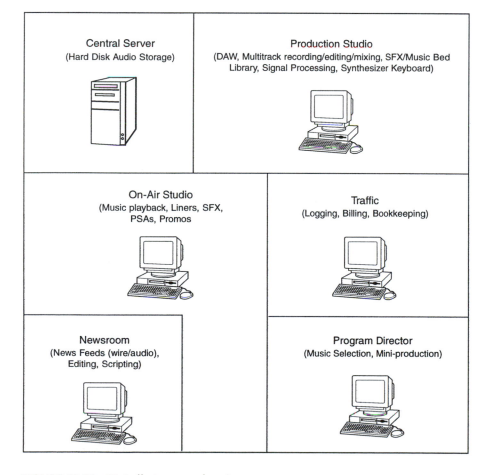

FIGURE 10.11 Digitally integrated station.

Another disadvantage of the digital equipment is noise—not audio noise, but noise from fans in computers or disk drives. If these have to be housed in a studio with live mics, there could be a problem. Another potential noise problem is "buzz" from a video monitor interfering with the speaker monitors. Shielding or creative placement might be necessary to make the digital studio workable in the normal studio environment.

10.10 THE FUTURE OF DIGITAL PRODUCTION

The future of digital technology, for those who embrace the concept, looks like more than just a very high-tech production studio. For example, most proponents of the DAW see it as a part of an all-digital radio station (see Figure 10.11). Not only can the DAW-based system produce a radio commercial, but it can store the commercial or several variations of it (all the time maintaining digital quality), play the commercial on-air by sending the signal to a digitized transmitter, and even send logging and bookkeeping information about the airing of the commercial to the appropriate station personnel. These systems can't be considered common in radio stations yet, but more and more radio stations are starting to use them.

10.11 DIGITAL DISTRIBUTION NETWORKS

Another way digital technology has impacted radio and radio production is the ability to deliver CD-quality audio to radio stations using **digital distribution networks**. Two ex-amples, Digital Generation Systems (DGS) and Digital Courier International (DCI), link radio stations with ad agencies, production houses, and record companies.

Under traditional analog techniques, radio spots were recorded onto audio tape masters, then dubbed to tape copies, and finally delivered by mail or courier to individual stations. The digital distribution networks feature PC-based servers in the ad agency, production house, or record company onto which audio files are loaded. This audio is sent via phone line to the network headquarters and then, with appropriate instructions, sent to receivers at client stations.

In addition to digital quality, these networks offer fast, instantaneous delivery to stations around the country and have been used to send commercials, newly released singles, and other production materials.

10.12 CONCLUSIONS

There seems to be little doubt that radio production has and will continue to benefit from the various developments of digital technology. Some industry experts feel that within a few years, digital technology will have entirely replaced analog. Others feel it will take considerably longer, but it's no longer a question of if this will happen, but rather when it will all come together. In any case, a basic understanding of digital technology and the operation of digital-based equipment will be more and more important in the future—and may soon be a prerequisite for success for the radio production person.

Self-Study

■ QUESTIONS

1. In which area is digital technology superior to analog technology?
   a) signal-to-noise ratio
   b) frequency response
   c) dynamic range
   d) all of the above

2. A standard IBM-compatible, Macintosh, or Amiga computer can be converted into a basic digital audio editor by adding appropriate software and _____.
   a) an ESP audio card
   b) an ESPN audio card
   c) a DSP audio card
   d) an ASP audio card

3. Digital cartridge machines use the same audio carts as analog machines but record in a digital format.
   a) true
   b) false

4. In general, a digital audio workstation's hard-disk drive will require _____ megabytes of memory for each stereo minute of audio the system is capable of handling.
   a) 5
   b) 10
   c) 15
   d) 20

5. Musical instruments (such as syntheziers or samplers) can interface to digital audio equipment through an electronic communications language known as _____.
   a) Bernoulli
   b) SETPM
   c) MINI
   d) MIDI

6. The MiniDisc recorder uses a _____ as its recording and storage medium.
   a) standard audio cassette
   b) 2.5-inch computer-type disk
   c) 3.5-inch computer-type disk
   d) DAT cassette

7. One of the main advantages of the digital audio workstation in radio production is _____.
   a) it's an inexpensive system to put together and maintain
   b) it's easy to operate
   c) it's capable of editing without cutting
   d) it offers better recording quality than the DAT

8. Digital audio workstations have the potential to introduce noise into the production process through two ways: cooling fan noise picked up by a mic and _____.
   a) video monitor interference picked up by a studio monitor speaker
   b) hiss picked up during the recording process
   c) clicks picked up at edit points during editing
   d) none of the above

9. The digital audio workstation has very little practical use outside the radio production studio.
   a) true
   b) false

10. Nondestructive editing is a feature of all desktop radio systems.
    a) true
    b) false

11. ATRAC (Adaptive Transform Acoustic Coding) is used for _____.
    a) auto-reverse of a DAW
    b) synchronization of musical instruments and audio equipment
    c) data compression
    d) recording audio on an IBM computer

12. Most digital equipment promises a longer interval between breakdowns than comparable analog equipment; however, when there is a technical problem, repair time for the digital equipment will probably also be longer.
    a) true
    b) false

13. Which disadvantage of the digital production studio would be easiest to argue against?
    a) can be difficult to learn digital equipment operation
    b) cost of digital equipment
    c) need for additional studio space for digital equipment
    d) sound of digital equipment

14. Which digital recording medium would have the longest storage time?
   a) MiniDisc
   b) Floptical Disc
   c) Hard Drive (2 gigabyte)
   d) Recordable CD

15. As a review of digital broadcast equipment, match the items in the top list (1, 2, 3 . . .) with the choices in the bottom list (s, v, c . . .), and then select the correct set of answers from the sequences shown in a, b, c, or d below.

   1. _____ analog
   2. _____ PC-based digital audio editor
   3. _____ digital cart machine
   4. _____ MIDI
   5. _____ DAW
   6. _____ DSP
   7. _____ SMPTE
   8. _____ MD

   s. a consumer level computer-based system that can store and manipulate sound in a variety of ways
   v. a continuously variable signal that represents the shape of a sound wave
   c. a computer card needed in order to edit audio with consumer-level computer equipment
   a. looks and operates like an analog cart machine
   m. an interface system used mainly to tie together musical instruments
   d. uses data compression system ATRAC
   i. an interface system that is useful for both audio and video
   f. an expensive but flexible nondestructive editing system

   a) 1.s 2.v 3.f 4.d 5.i 6.a 7.m 8.c
   b) 1.v 2.s 3.a 4.m 5.c 6.f 7.i 8.d
   c) 1.v 2.s 3.a 4.i 5.f 6.c 7.m 8.d
   d) 1.v 2.s 3.a 4.m 5.f 6.c 7.i 8.d

■ ANSWERS

If You Answered A:

1a. You're partially correct, but this is not the best response. (Reread 10.1 and 10.8.)
2a. No. You didn't "sense" the correct answer. (Reread 10.5.)
3a. Wrong. Some digital cart machines use a 3.5-inch computer disk; others use different types of disks as the recording medium. (Reread 10.2.)
4a. No. Hint: this would work for a mono signal. (Reread 10.6.)
5a. Wrong. A Bernoulli cart is used in conjunction with a digital cart, but it isn't a communications language. (Reread 10.2 and 10.7.)
6a. No. The audio cassette can only record in an analog cassette recorder and play in the same or a DCC recorder. (Reread 10.3.)
7a. No. DAWs are relatively expensive. (Reread 10.6, 10.8, and 10.9.)
8a. Correct. This could be a problem with incorrectly positioned monitors.
9a. No. (Reread 10.10.)
10a. No. Some desktop radio systems have a destructive edit mode and actually delete the audio that has been edited, but usually only after a preview of the edit. (Reread 10.5.)
11a. Wrong. This isn't a feature found on DAWs. (Reread 10.4 and 10.6.)
12a. No. Only half of this statement is true. While digital equipment will probably break down less than analog equipment, repair for digital equipment also promises to be shorter in many cases. (Reread 10.8.)
13a. No. Some digital equipment isn't very user friendly and has a considerable learning curve. (Reread 10.9.)
14a. No. MDs can record to a length of 74 minutes. (Reread 10.3 and 10.6.)
15a. You made many mistakes. (Reread the entire chapter.)

If You Answered B:

1b. You're partially correct, but this is not the best response. (Reread 10.1 and 10.8.)

2b. No. You might have just taken a "sporting" guess to get this answer. (Reread 10.5.)

3b. Right. There are several types of disks that are used as the recording medium for the digital cart machine, but none are the same as the analog carts.

4b. Yes. This is the correct answer.

5b. Wrong. SMPTE might be a possible answer, but there is no communication language called SETPM. (Reread 10.7.)

6b. Yes. The MiniDisc uses a 2.5-inch computer-type disk.

7b. No. DAWs are complicated and require some time to learn to operate proficiently. (Reread 10.6, 10.8, and 10.9.)

8b. Wrong. Digital recording will add virtually no hiss. (Reread 10.8 and 10.9.)

9b. Yes. DAWs could well become the heart of a completely digital radio station.

10b. Yes. There are desktop radio systems that have both destructive and nondestructive edit modes.

11b. Wrong. You're thinking of MIDI if you selected this answer. (Reread 10.4 and 10.7.)

12b. Yes. Only part of this statement is true. Digital equipment repair time should be shorter than analog.

13b. No. While some digital equipment is costly, as more and more digital equipment hits the market, this is less and less of a disadvantage. However, there is a better answer. (Reread 10.9.)

14b. No. A floptical disc can only record about 5 minutes of digital quality audio. (Reread 10.2 and 10.6.)

15b. No. You are confusing some of the abbreviations. (Reread 10.5 and 10.6.)

If You Answered C:

1c. You're partially correct, but this is not the best response. (Reread 10.1 and 10.8.)

2c. Correct. A digital signal-processor audio card is part of the basic digital audio editor.

4c. No. (Reread 10.6.)

5c. Wrong. You're probably confusing this with the MiniDisc. (Reread 10.3 and 10.7.)

6c. No. You're close, but this size disk is used in another digital cart machine. (Reread 10.2 and 10.3.)

7c. Yes. DAWs offer nondestructive editing capability.

8c. Wrong. DAW electronic editing is noiseless. (Reread 10.9.)

11c. Correct. This is SONY's data reduction scheme to put 74 minutes of recording time on a tiny computer-type disk.

13c. No. It's true some digital equipment is smaller than comparable analog equipment, but usually additional studio space is required for keyboards, computer monitors, etc. (Reread 10.9.)

14c. Yes. Most likely a hard drive will have a great deal of recording time, but remember it takes about 10 MB to record each stereo minute.

15c. No. You confused the interface systems. (Reread 10.7.)

If You Answered D:

1d. Correct. Digital technology offers all these technical improvements over analog.

2d. No. You got "bitten" on this answer. (Reread 10.5.)

4d. No. (Reread 10.6.)

5d. Right. MIDI (musical instrument digital interface) is the correct response.

6d. No. DAT cassettes are only used in DAT recorders/players. (Reread 10.3.)

7d. No. Both DATs and DAWs record with the same digital quality. (Reread 10.6, 10.8, and 10.9.)

8d. Wrong. There is a correct response, but this isn't it. (Reread 10.9.)

11d. Wrong. You're thinking of a DSP card if you selected this answer. (Reread 10.4 and 10.5.)

13d. Yes. This is the weakest disadvantage of digital equipment because in most ways digital audio is obviously superior in quality to analog and in other ways it's a highly subjective judgment.

14d. No. You learned in a previous chapter that recordable CDs can have a 74-minute recording length. (Reread 10.6.)

15d. Right. You have now completed this chapter.

Projects

■ PROJECT 1

Write a report that compares various types of digital cartridge recorders.

Purpose

To allow you to further investigate newer technology in the area of digital electronics.

Advice, Cautions, and Background

1. Prepare a list of questions that you want answered before you begin your research.
2. This project requires obtaining material from various manufacturers, so give yourself plenty of time to complete the project.

How to Do the Project

1. Develop some questions that you want answered about floptical disk, hard disk, and the MiniDisc recorder, for example:
   a. How would you describe the physical configuration of each technology?
   b. What exactly can each do in the area of radio production?
   c. How much do they cost?
   d. What problems are associated with these technologies?
   e. What recording medium does each technology use?
2. Gather information by writing directly to some of the manufacturers. (Some addresses are provided below)
3. If your facility has this equipment or you know of a station in your area that does, try to talk with someone who uses this equipment on a regular basis to further research the technology.
4. Organize your material into an informative report that compares the technologies. It should be several pages long. Write your name and DIGITAL CARTRIDGE RECORDERS COMPARISON on a title page.
5. Turn in the report to your instructor for credit for this project.

MiniDisc
Denon Electronics
222 New Road
Parsippany, NJ 07054

Hard Disk Recorder
360 Systems
5321 Sterling Center Drive
Westlake Village, CA 91361

Floptical Disk Recorder
Fidelipac Corporation
97 Foster Road
Moorestown, NJ 08057

■ PROJECT 2

Using a digital audio editor or digital workstation, build a short music bed.

Purpose

To give you experience with digitally based editing.

Advice, Cautions, and Background

1. Obviously, you can't do this assignment unless you have the hardware and software to complete it available to you.
2. CD music that you can use is provided on the CD that accompanies this text, but if you prefer, you can use your own music from some other source—a DAT, an analog tape, a record. Keep in mind that a CD or music on a DAT tape will already be in digital form, whereas music from a record or analog tape will be in analog form.
3. You'll be laying the same 5 seconds of music down three times, so select some music that will lend itself to this. Find 5 seconds at the beginning or in the middle of some piece that won't sound like it has been cut off abruptly.
4. The project is written to be 15 seconds long, but if you have a system that can store a great deal of audio, feel free to make a music bed 30 or 60 seconds long.

How to Do the Project

1. Select music from the CD provided or from some other source and transfer about 10–20 seconds of it into your computer.
2. Activate your computer editing program.
3. Select 5 seconds of music from the 10–20 seconds, and lay it down on one track made available through your computer editing program.
4. Lay the same 5 seconds of material down again right next to the first 5 seconds, and then lay it down one more time right after the second 5 seconds.
5. Listen to it and improve upon it so that it could be used as a music bed under someone talking about a commercial product. You may want to change volume or fade it in or out or tighten it a bit. What you can do and exactly how you do it will depend on the computer program that you're using.
6. Record from the computer to an audio tape (DAT or analog). Label this tape with your name and DIGITAL EDITED MUSIC BED. Give the tape to your instructor to receive credit for this project.

CHAPTER ELEVEN

Multitrack Production Techniques

Information

11.1 INTRODUCTION

While a few larger market stations had multitrack reel-to-reel recorders in their production studios, for years most of the radio production work was done in either mono or two-track stereo. Now, as more and more digital equipment finds its way into the production studio, more and more of the production that's done will be multitrack. As noted in the last chapter, even basic, inexpensive desktop radio systems work in a multitrack mode. The biggest advantage of multitrack recording is that it opens the door to more creative production work. In this chapter you'll learn some of the basic techniques of multitrack production; however, these should merely suggest ways to begin to approach your future productions.

11.2 THE MULTITRACK RECORDER

In addition to the production equipment already mentioned in this text, some production facilities have a multitrack recorder like the reel-to-reel shown in Figure 11.1. Any machine that can record more than one track is a multitrack recorder. However, in radio production, multitrack generally means recorders that have four, eight, or more tracks.

Reel-to-reel recorders aren't the only type of multitrack system. The digital recorder shown in Figure 11.2 is a multitrack system that employs S-VHS tape as the recording and storage medium. As previously noted, most digital audio workstations, like the ones shown in Figures 10.9 and 10.10, also provide multitrack capability utilizing their software programs and hard disc drive.

The multitrack recorder is really part of a multitrack system that contains these basic elements: an audio mixer, multitrack recorder, signal processors, and input sources and output devices. Sometimes several elements are combined; for example, a multichannel mixer with equalization capability and a multitrack recorder with noise reduction capability may be combined in a single unit.

The audio mixer may be the production room board or a dedicated board that is associated only with the multitrack recorder. In either case, it will need a good number of inputs and outputs—while a microphone is usually assigned to just one channel, any stereo source (such as a CD player) must be assigned to two channels. In general, the mixer should have at least one input/output channel for each track of the multitrack recorder. In other words, an eight-track recorder should have a mixer with at least eight channels associated with it. In addition, the mixer must be able to input each of the tracks of the multitrack recorder so that monitoring and mixing can take place. One arrangement might put all the normal inputs (mics, CDs, etc.) on "program" of the mixer and all the tape inputs on "audition" or "auxiliary."

Signal processing usually includes equalization, noise reduction, and reverb effects at the least, and they can be much more elaborate. Some signal processing is usually associated with individual channels of the mixer; for example, each channel may have some EQ adjustment. In other cases, the ouput of a channel (such as a mic) may be sent to a signal-processing device (such as a digital reverb), and then the output of the signal-processing device is sent back to another input on the mixer, where the "processed" signal can be routed to the multitrack recorder.

The input sources are normal production inputs, such as microphones, CD players, or audio cartridge players, and the output devices are either headphones or monitor speakers that allow for listening to the mix of all the tracks.

The multitrack recorder operates like any reel-to-reel recorder as discussed in Chapter Five. Many multitrack recorders employed in radio production use audio tape that is ½-inch or 1-inch wide; however, there are some that use standard ¼-inch tape, cassette tape, various videotape formats, and hard disc drives. Reviewing the tape track configuration in Chapter Five shows that a four-track recorder would have four separate tracks being recorded on the tape in one direction. Each track can be recorded and played back separately so that you can record one track while listening to another. This is a primary advantage of multitrack recording.

FIGURE 11.1 Multitrack recorder (4-track). (Courtesy of Studer Revox America, Inc.)

11.3 OVERDUBBING TRACKS

One of the most basic techniques used in multitrack work is **overdubbing,** or the process of adding new tracks to existing tracks. For example, in the most basic form, you would record a stereo music bed on tracks 1 and 2 of a multitrack recorder. Then you recue the recorder to the beginning of the music bed. Now, while you listen to the playback of the music bed (usually through headphones) on tracks 1 and 2, you record a vocal on track 3. All multitrack recorders have a sel sync feature that lets you listen to one track and record on a different track in synchronization. Overdubbing allows you to build a production in "layers" because you don't have to record everything at the same time.

11.4 PUNCHING IN TRACKS

If you've recorded a vocal on one track, but a portion of the vocal contains a mistake, a multitrack recording technique known as the **punch in** allows you to record over just the part that contains the mistake, and leave the rest of the track undisturbed. Once you've completed the initial recording, recue the track to the beginning. The multitrack recorder will allow you to monitor both the playback of the track and the live rerecording. It's important that you don't change volume levels or mic position of the original setup so that the rerecording is consistent with the original recording. The multitrack recorder also allows you to put the recorder into "record" mode by pressing a record button for the individual track you want to record on or putting that track into a "record ready" mode, then pressing a master record button on the multitrack. The technique is a punch in because you playback the track to the point where you

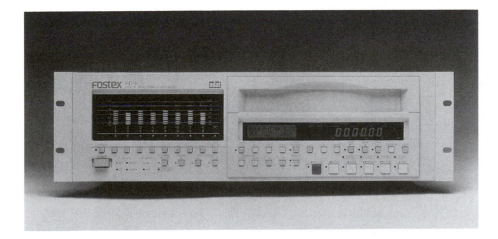

FIGURE 11.2 Digital multitrack recorder (S-VHS tape recording medium). (Courtesy of Fostex Corporation of America)

want to rerecord, then punch in the record button and re-cord over the portion of the track that you wish to correct. You punch out of record mode at an appropriate point after you've fixed the mistake. Essentially you're doing an "edit" on the fly with a punch in; in fact, it's also known as an **insert edit**. You might need to practice a few times, so you know exactly where you want to punch in and punch out. It makes sense to choose logical spots, such as the end of a sentence or at a pause between words, otherwise the punch in may be noticeable.

11.5 BOUNCING TRACKS

Bouncing, or **Ping-Ponging**, tracks is the process of com-bining two or more tracks on a multitrack recorder and re-recording and transferring them to another vacant track of the same recorder. If you were building a music bed for ex-ample, record drums on track 1 and a bass on track 2. Then bounce those tracks to track 3. Now you have track 1 and track 2 free, and you can record stereo strings on those two tracks, and, if you want, bounce all three tracks to track 4.

Bouncing is most often used when the multitrack recorder has a limited number of tracks, such as four, and you're do-ing complex production work. In other words, it's a tech-nique to increase the number of tracks that you can utilize. However, bouncing tracks brings with it two disadvantages the production person must be aware of. First, once several tracks are combined, they can't be uncombined. You can't change the balance of the mixed tracks, and if you equalize or otherwise process the track, it impacts all elements on the track. Second, each "bounce" causes a degradation of the sound quality of the track if the recorder is analog. It's just like dubbing from one tape to another; once you've bounced to one track, that track is a second generation re-cording. If you bounce that track again, it's become a third generation recording with resulting lower audio quality. Careful planning can help keep the number of tracks that you must bounce to a minimum.

Live bouncing is a similar technique, but at the same time that you're mixing several tracks onto a vacant track, you add a live track to the mix. Again, it's a technique to get a maximum number of sources onto a minimum number of tracks.

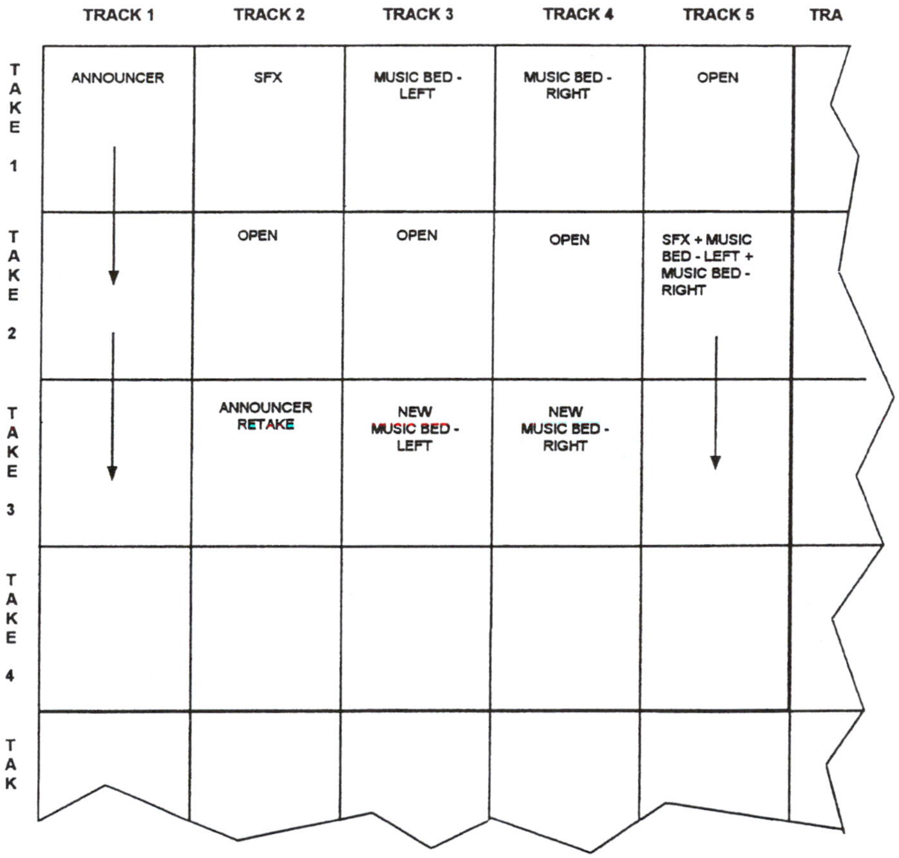

	TRACK 1	TRACK 2	TRACK 3	TRACK 4	TRACK 5	TRA
TAKE 1	ANNOUNCER	SFX	MUSIC BED - LEFT	MUSIC BED - RIGHT	OPEN	
TAKE 2		OPEN	OPEN	OPEN	SFX + MUSIC BED - LEFT + MUSIC BED - RIGHT	
TAKE 3		ANNOUNCER RETAKE	NEW MUSIC BED - LEFT	NEW MUSIC BED - RIGHT		
TAKE 4						
TAK						

FIGURE 11.3 Sample track sheet.

11.6 TRACK SHEETS

As you can tell just by reading about the techniques above, multitrack recording can become complicated. Good production practice dictates that you keep notes of what material is recorded on which track. While there is no standard form, Figure 11.3 shows one possible **track sheet**, or format, for keeping notes regarding a multitrack production. The top of the page indicates the number of tracks the recorder has, the edge of the page indicates the various takes, and the boxes show what was put on each track during each take. As shown, during the first take, an announcer vocal was put on track 1, a sound effect was put on track 2, a music bed was put on tracks 3 and 4, and track 5 was left open. This doesn't mean that these elements were all put on the tape at the same time; most likely they were put down separately, but these were the first elements assigned to those tracks. During the second take, the announcer vocal was not changed, but the sound effect on track 2 and music bed on tracks 3 and 4 were all bounced to track 5. On the third take, tracks 1 and 5 did not change; however, a new music bed was put on tracks 3 and 4, and the announcer vocal was rerecorded on track 2.

You could also put any signal processing settings, music bed cuts used, or special notes regarding the production on the track sheet.

11.7 MIXING TO STEREO/MONO COMPATIBILITY

Regardless of how many tracks you work with in a multitrack production, you will ultimately mix the production down into a stereo mix with just a left and right channel. Certain tracks from the multitrack master may be panned to the right channel, some tracks may be panned to the left channel, and some tracks may be balanced to both left and right. This is where the production person needs a good monitoring system and some good judgment to create just the right mix. Make sure that you listen to your final sound in both stereo and mono. Most likely you're working in stereo and hearing the final mix in stereo; however, you need to hear the same mix in mono because sometimes a mix sounds great in stereo, but not in mono. It's possible that portions of the audio signal will be out of phase when the left and right channels are combined for the mono signal, causing a cancellation and diminishing of the sound at different spots. A fine tuning of some track equalization, a slight volume change, or minor panning adjustment will usually correct the mono signal, but it's always important to make sure the stereo and mono signals are compatible.

11.8 THE MULTITRACK COMMERCIAL SPOT

There isn't a "standard" way to record a commercial using multitrack techniques, but let's look at the production of a spot that includes two announcer voices, background music bed, and two sound effects:

a. music bed at full volume for a few seconds
b. music bed fades under and holds
c. announcer #1 begins voice-over on top of music bed
d. announcer #2 trades lines with announcer #1
e. sound effects added at appropriate points
f. music bed brought up to full volume at end of voice-over
g. music bed ends cold

Assuming a production studio setup with an eight-track recorder, refer to Figure 11.4 to help visualize what is happening during this "typical" production.

Each sound source (mic, CD, etc.) should be assigned a channel on the mixer and corresponding track on the recorder. Let's put announcer #1 on track 1, announcer #2 on track 2, a CD sound effect on tracks 3 and 4, a CD music bed on tracks 5 and 6, and a cart sound effect on tracks 7 and 8. The first track you record is extremely important because it's the reference for all the other tracks. In radio production, the music bed or vocal track is often put down first.

In this case, the music bed is a preproduced and timed music bed from a CD production library. We just have to dub it from the CD onto our recorder. Of course, the music bed could have come from another source—for example, a music bed directly from a record or perhaps a music bed that was edited on reel-to-reel to be exactly the length of the spot. Now set the multitrack recorder to record on tracks 5 and 6 only (the music bed). Dub the entire music bed at full volume onto the recorder, even though you'll be fading it under the announcers. The various levels can be balanced during the final mix.

Now put the recorder tracks 5 and 6 into a "safe" and "sync" mode. This allows you to overdub on the other tracks at the same time that you monitor the music bed tracks. Set tracks 1 and 2 into the record position. Since the announcers "trade" lines throughout the commercial, you should record them at the same time. Rewind the tape to the beginning of the music bed. While monitoring tracks 5, 6, 1, and 2, record announcers #1 and #2. You should fade the music at the appropriate place, so the announcers get the proper feel reading the spot, but remember the actual "blend" of all the elements will take place later. At this point, if the vocal tracks came out OK, the announcers could leave, and a production person would finish the spot. That makes sense if the announcers are "high-priced" talent and that's why many multitrack productions are started by putting down the vocal tracks first. Suppose announcer #2 misspoke the final line in the spot. Depending on the complexity of the spot, it would be possible to rewind the tape, set up tracks 5, 6, and 1 to play, and have announcer #2 rerecord track 2 to correct the mistake with a punch-in.

By additional overdubbing, the various sound effects would be added by playing back the tracks that were previously recorded and recording only those tracks that were

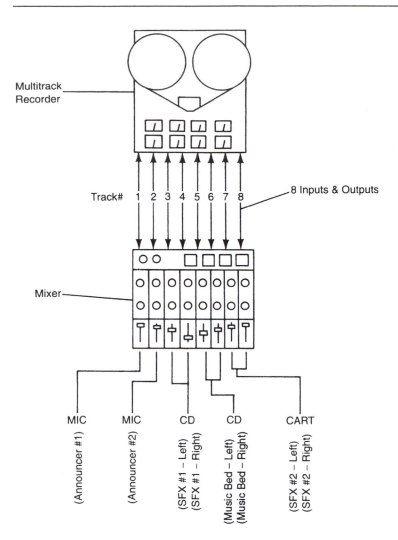

FIGURE 11.4 Multitrack recording.

assigned to sound effects, doing one effect at a time. Unlike the basic commercial production, if you miscue a sound effect, you don't have to reset every element and start all over again. You would merely rerecord the track that had the mistake on it.

Once you have all eight tracks recorded, you can begin to mix down the final spot. Since most commercials are played on a stereo format, they are usually mixed down to a two-track (stereo) reel-to-reel master. You'll probably play back all eight tracks several times now (through the mixer, monitoring the multitrack recorder outputs) to adjust the levels. For this commercial, you can get the announcer and sound effect levels balanced and leave them set, but you'll have to fade the music bed up and down manually at the appropriate points as you record onto the two-track. With a more complex spot, you might have to manipulate more than one fader during the final mix—and that's where practice and experience come in to produce the perfect spot!

11.9 VOICE DOUBLING, CHORUSING, AND STACKING

Multitrack techniques open the door to many special effects that can turn a basic production into a creative masterpiece. Voice doubling, chorusing, and stacking are three forms of overdubbing that can give your next production a unique sound. **Voice doubling** is exactly what the name implies. Record your voice on track 1. Then recue the multitrack recorder. Now while monitoring track 1, record your voice again on track 2. Even though you read the same script, it's impossible to record both tracks exactly alike. The effect, also known as "voice dubbing," will be closer to two people reading the same script at the same time. **Chorusing** is taking voice doubling one step further. Record at least two additional tracks in sync with the original track, creating a "chorus" effect. The more additional tracks you record, the larger your "chorus" will sound. It's a way for one announcer to become a group of announcers. **Stacking** is a similar multitrack technique in which an announcer *sings* harmony to a previously recorded track.

11.10 DOVETAILING

One announcer can appear to be two different announcers trading lines using a technique referred to as **dovetailing**. On track 1, the announcer records the *odd* lines of the script. While doing so, the even lines of the script are also mentally read. This leaves space between the odd lines on track 1. Now, record on track 2 the *even* lines of the script while mentally reading the odd lines to keep the timing correct. When both tracks are played back, you get the effect of two announcers reading a dialogue script. Obviously you must slightly change one of the voices so that it sounds like two different announcers, and you'll probably need to practice a few times to get the timing correct. Remember, you usually "mentally" read lines slightly faster than you do when you actually speak them. It can take a little time to perfect this technique, but it can be well worth the effort.

11.11 SLAP-BACK ECHO

To produce a slap-back echo with a multitrack recorder, record your voice on track 1. Now, dub track 1 to another track, but without synchronization. Play back both tracks, and you'll get a unique echo effect. If you do the recording at a higher recorder speed, then you'll get a "closer" echo effect.

11.12 DIGITAL MULTITRACK

We've already made the point that, from the basic desktop editor to the complex workstation, most digital systems have multitrack capability. Figure 11.5 shows a typical screen from one such system. In this view, the last four tracks of the system are shown with some of the controls associated with those tracks. At the right of the screen, the audio data in each track is shown as a block known as a "**region**." Regions can easily be moved, copied, or deleted. This information can be displayed in different sizes with a "zoom" control and can also be shown as the actual waveform of the audio as we see here. To the left of the screen are virtual faders and pan controls for each track, which allow for level setting and balance control during the mix of the production. All the multitrack techniques mentioned previously in this chapter are utilized with digital systems, but

you may find that actually manipulating the audio is easier with a digital system. For example, to move a region, a "select" mode is engaged, and then a mouse click selects the region you want to move. You can now "drag and drop" that region from one track to another, from one part of a track to another, or copy or delete the region. Remember, when you copy something in digital form, the sound doesn't degrade as it does with dubbed analog generations.

From a simple combining of a music bed and voice-over track to a complex production using hundreds of regions, digital multitrack offers ease of operation and creative possibilities to the modern production person.

11.13 CONCLUSIONS

As we've noted, much of the new digital equipment that's making its way into the radio production studio has multitrack capability. The modern production person is going to have to be skilled in working in the multitrack environment to be successful. While this chapter is only designed to give you some basic familiarity with the equipment and techniques of multitrack production, it should serve as a good starting point. Like most everything about radio production, the real learning comes when you go into the studio and try out different concepts. You'll really become skilled in multitrack production work only if you spend time in the studio to understand the creative possibilities it offers.

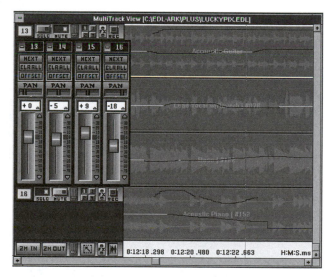

FIGURE 11.5 Digital editor multitrack screen. (Courtesy of Innovative Quality Software)

Self-Study

■ QUESTIONS

1. From the following number of tracks, how many tracks is a multitrack recorder found in the radio production studio likely to have?
   a) 1
   b) 2
   c) 4
   d) 6

2. When two or more tracks of a multitrack recording are combined and rerecorded on another vacant track, this is known as _____ tracks.
   a) overdubbing
   b) bouncing
   c) punching-in
   d) combo-cording

3. Which of the following is *not* part of a multitrack recorder system?
   a) audio mixer
   b) multitrack recorder
   c) DSP audio card
   d) microphone

4. Which type of tape is *not* used as a recording medium in multitrack recording systems?
   a) cassette tape
   b) video tape
   c) reel-to-reel tape
   d) all of these are multitrack recording media

5. In multitrack recording, the process of adding new tracks to existing tracks is known as _____.
   a) bouncing tracks
   b) overdubbing tracks
   c) punching in tracks
   d) Ping-Ponging tracks

6. One announcer can appear to be two different announcers reading a dialogue script using a multitrack recording technique known as _____.
   a) voice doubling
   b) chorusing
   c) dovetailing
   d) stacking

7. The main reason announcer voices are often recorded first on a multitrack recording is because vocal sources are usually assigned to track 1 and track 2.
   a) true
   b) false

8. Which of the following statements is *not* associated with bouncing tracks during a multitrack recording?
   a) you can combine two or more tracks on a vacant track of the same recorder
   b) several combined tracks can't be uncombined
   c) several bounces can cause degradation of sound quality
   d) you are rerecording a portion of a track

9. A track sheet is a manufacturer specification sheet that lists the number of tracks that a multitrack recorder has.
   a) true
   b) false

10. Multitrack productions are usually mixed down to a two-track stereo master. If the stereo master is slightly out-of-phase when combined into a mono signal, it can be corrected by all but which of these adjustments to one of the stereo tracks?
    a) change track equalization
    b) move track to a vacant track
    c) change track volume
    d) change track panning

11. During multitrack recording, which technique allows you to rerecord just a portion of a track to correct a mistake while leaving the rest of the track undisturbed?
    a) punching in
    b) overdubbing
    c) voice doubling
    d) bouncing

12. Bouncing tracks on a multitrack recorder is also known as _____.
    a) chorusing
    b) overdubbing
    c) Ping-Ponging
    d) dovetailing

13. On a 4-track multitrack tape recorder _____.
    a) all the tracks go the same direction
    b) two tracks go left and two go right
    c) only tracks 1 and 3 can be used
    d) you can record 8 tracks of sound

14. What is one advantage of digital multitrack over analog multitrack?
    a) digtial multitracks have six tracks, while analog can have no more than four tracks
    b) the sel sync feature is in better synchronization on digital than analog
    c) you can punch in on digital, and you can't on analog
    d) you can manipulate sounds easier with digital because you can move them with the mouse

15. As a review of multitrack production techniques, match the items in the top list (1, 2, 3 . . .) with the choices in the bottom list (c, d, p . . .), and then select the correct set of answers from the sequences shown in a, b, c, or d below.

    1. _____ stacking
    2. _____ overdubbing
    3. _____ Ping-Ponging
    4. _____ track sheet
    5. _____ multitrack system
    6. _____ chorusing
    7. _____ dovetailing
    8. _____ punching in
    9. _____ slap-back echo

    c. consists of an audio mixer, multitrack recorder, signal processor, and input and output devices
    d. a device for keeping notes on what you recorded on which tracks
    p. a process that involves dubbing from one track to another without synchronization
    a. adding new tracks to existing tracks
    r. recording over (on the fly) a small part of something you recorded before
    m. a method to allow one person to record the dialogue of two people
    t. recording at least two additional tracks almost in sync with an original track

s. singing harmony to a previously recorded track

b. also known as bouncing

a) 1.s 2.a 3.b 4.d 5.c 6.t 7.m 8.r 9.p

b) 1.b 2.r 3.s 4.c 5.d 6.m 7.t 8.p 9.a

c) 1.b 2.a 3.s 4.d 5.c 6.m 7.t 8.r 9.p

d) 1.s 2.a 3.b 4.c 5.d 6.t 7.m 8.r 9.p

■ ANSWERS

If You Answered A:

1a. No. One would not be "multi." (Reread 11.2.)

2a. Wrong. Overdubbing is the normal technique for multitrack recording. (Reread 11.3 and 11.5.)

3a. No. This is part of the multitrack setup. (Reread 11.2.)

4a. No. There are multitrack recorders that use cassette tapes. (Reread 11.2.)

5a. No. This is another multitrack recording technique. (Reread 11.3 and 11.5.)

6a. Wrong. While this is a "voice doubling" technique, it's another multitrack trick. (Reread 11.9 and 11.10.)

7a. No. It really doesn't matter what sources are assigned to what tracks. (Reread 11.8.)

8a. No. This is exactly what bouncing tracks is. (Reread 11.4 and 11.5.)

9a. No. A track sheet is a way of keeping notes during a multitrack production. It usually lists track numbers and what is recorded on each track. (Reread 11.6.)

10a. Wrong. This will often correct the mono signal. (Reread 11.7.)

11a. Yes. The punch-in is also known as an insert edit because you can just rerecord a portion of a track.

12a. No. This is a different multitrack technique. (Reread 11.5 and 11.9.)

13a. Yes. All the tracks must go in the same direction, so you can listen and record at the same time.

14a. No. Both can have a varying number of tracks. (Reread 11.2 and 11.12.)

15a. Right. You have now completed this chapter.

If You Answered B:

1b. You're close because two-track recorders are usually found in the production studio, and technically they are multitrack, but this is not the best answer. (Reread 11.2 to find out why.)

2b. Correct. Bouncing tracks or Ping-Ponging tracks is the multitrack technique described here.

3b. No. This is an obvious part of the multitrack setup. (Reread 11.2.)

4b. No. There are multitrack recorders that use S-VHS video tape. (Reread 11.2.)

5b. Yes. This is a primary advantage of multitrack recording and one of the most basic techniques used in multitrack production.

6b. Wrong. While this is a technique for adding voices, it's another multitrack technique. (Reread 11.9 and 11.10.)

7b. Yes. It doesn't really matter what sources are assigned to what tracks, but announcers are often recorded first because it's easier to lay down other tracks to the vocals and "high-priced" talent can be finished with their part of a production once their tracks are recorded, even if the entire spot isn't completed.

8b. No. This is a minor disadvantage of bouncing tracks. (Reread 11.4 and 11.5.)

9b. Yes. This is not a correct statement.

10b. Correct. Just moving a track's location won't help.

11b. No. This technique allows you to add additional tracks to a production. (Reread 11.3 and 11.4.)

12b. No. This is a different multitrack technique. (Reread 11.3 and 11.5.)

13b. No. This would be very confusing. (Reread 11.2.)

14b. No. You are way off. (Reread 11.3 and 11.12.)

15b. You made many mistakes. (Reread the entire chapter.)

If You Answered C:

1c. Yes. Of the answers offered, this is the most likely configuration of a multitrack recorder in the production studio.

2c. Wrong. To punch-in is a technique that allows you to record a segment of a track without affecting the material before or after that segment. (Reread 11.4 and 11.5.)

3c. Correct. A DSP audio card is part of a simple audio editing system, and while the system may have multitrack recording capability it's different from a multitrack recorder system.

4c. No. Many multitrack recorders use reel-to-reel audio tape. (Reread 11.2.)
5c. No. This is another multitrack recording technique. (Reread 11.3 and 11.4.)
6c. Correct. By reading even lines of a script on one track and then odd lines on another track, one announcer can sound like he or she is talking to another with this technique.
8c. No. This is a minor disadvantage of bouncing tracks. (Reread 11.4 and 11.5.)
10c. Wrong. This will often correct the mono signal. (Reread 11.7.)
11c. No. This technique is a form of overdubbing that allows you to "voice double." (Reread 11.4 and 11.9.)
12c. Yes. This is another term for bouncing tracks.
13c. No. This would limit the usefulness of the recorder. (Reread 11.2.)
14c. No. You can punch-in on analog. (Reread 11.4 and 11.12.)
15c. You are confused about four things. (Reread 11.5, 11.6, 11.9, and 11.10.)

If You Answered D:

1d. No. This is not a common number of tracks for a radio production multitrack recorder. (Reread 11.2.)
2d. Wrong. There is no such thing. (Reread 11.5.)
3d. No. This is a possible input source for a multitrack recorder system. (Reread 11.2.)
4d. Yes. This is the correct answer.
5d. No. This is another multitrack recording technique. (Reread 11.3 and 11.5.)
6d. Wrong. While this is a technique for adding voices, it's another multitrack technique. (Reread 11.9 and 11.10.)
8d. Yes. This is not accomplished by bouncing tracks, but by a technique known as the punch-in.
10d. Wrong. This will often correct the mono signal. (Reread 11.7.)
11d. No. This technique allows you to combine tracks and transfer them to a vacant track. (Reread 11.4 and 11.5.)
12d. No. This is a different multitrack technique. (Reread 11.5 and 11.10.)
13d. No. It's only a four-track recorder. (Reread 11.2.)
14d. Yes. A few clicks of the mouse can easily move sounds with a digital multitrack system.
15d. You made two small mistakes. (Reread 11.2 and 11.6.)

Projects

■ PROJECT 1

Record a creative dialogue 30-second or 60-second commercial using dovetailing technique.

Purpose

To develop your skill in creating a commercial using multitrack techniques and to train you to work within a specific time limit.

Advice, Cautions, and Background

1. This project assumes you have a multitrack recorder in your studio; however, it can also be accomplished with a two-track recorder.
2. The commercial must be exactly 30 seconds or 60 seconds.
3. You can write the commercial either for an actual product or for something you make up.
3. Be creative in terms of the idea, the wording, and the special effects, but remember: It must utilize dialogue of two people.
4. Keep in mind that the commercial is to sell a product or idea, and the creativity should enhance the subject and not detract from it.

How to Do the Project

1. Think up an idea for the commercial. Make sure your idea is producible; i.e., that you have all the music, effects, and equipment necessary to do it.
2. Write the commercial, making sure it is 30 or 60 seconds.
3. Practice reading one person's dialogue while mentally leaving room for the other person's dialogue.
4. When you think you have this part of the dovetailing technique down, record the first person's dialogue on one of the tracks.
5. Change the tone of your voice, playback the track you just recorded, and practice reading the second person's dialogue.
6. When you feel ready, record the second dialogue on a different track.
7. Add music and effects (if you are using them), and then mix the entire commercial down to two-track stereo.
8. Listen to the finished commercial. If it's good enough, prepare it for the instructor. If not, redo whatever needs to be redone.
9. On the tape write your name and DOVETAIL COMMERCIAL PRODUCTION, and turn it in to your instructor to receive credit for this project.

■ PROJECT 2

Use a multitrack recorder to bounce tracks, and keep a track sheet to show what you have done.

Purpose

To enable you to learn the commonly used multitrack technique of bouncing and to allow you to learn how to keep a track sheet.

Advice, Cautions, and Background

1. This project assumes you have a multitrack recorder in your studio. If you don't have one, but you have a digital audio workstation, you can use that instead.
2. The project also assumes that you have the multitrack recorder in a system that has an audio board, a microphone, and something (such as a CD player) that can play music.
3. Make one track music and one voice so that you can get practice mixing these two types of sounds.
4. You may find music appropriate for this project on the audio CD that came with this text.

How to Do the Project

1. Record music on track 1.
2. Record voice on track 2. If you make a mistake, rerecord your voice on this track.
3. Mix tracks 1 and 2 onto track 3 so that the voice is easily distinguishable.
4. As you're doing your recording, keep a track sheet similar to the one depicted in section 11.6 that shows what you did.
5. Put your name on the multitrack tape and the track sheet.
6. If you've used a DAW, you'll need to dub your work onto an audio tape. If so, dub each track separately.
7. Label the tape and track sheet with your name and BOUNCING TRACKS.
8. Turn in the tape and track sheet to your instructor to receive credit for this project.

■ PROJECT 3

Use a multitrack recorder to undertake chorusing.

Purpose

To teach you the multitrack technique of chorusing and to further your skills with multitrack bouncing.

Advice, Cautions, and Background

1. This project assumes you have a four-track recorder in your studio. If the one you have has more than four tracks, only use four of the tracks.
2. You can use any dialogue you want, but the first line of "Mary Had a Little Lamb" works well.
3. You will use track 1 twice when doing this project.

How to Do the Project

1. Record your selected dialogue on track 1.
2. While playing back track 1, record the dialogue again on track 2.
3. Mix tracks 1 and 2 onto track 3.
4. While listening to track 3, record the dialogue again on track 1.
5. Mix tracks 3 and 1 onto track 4.
6. Listen to track 4. It should sound like a chorus of four voices.
7. Turn in the tape, with something on all four tracks, to your instructor to receive credit for this project. Put your name and CHORUSING PRODUCTION on the tape.

CHAPTER TWELVE

Production Tips, Tricks, and Techniques

Information

12.1 INTRODUCTION

Throughout this text, you have been introduced to various production principles and techniques: slip cueing a record, preventing multiple-microphone interference, and editing audio tape, to mention a few. In this chapter we'll provide some additional production tips, tricks, and techniques that you might find useful in the radio production studio. Some of these production techniques are valid for many different production situations and should be useful every time you go into the broadcast studio. Others are specialized tricks and techniques that may prove helpful only on occasion. While very creative production work can be accomplished through multitrack recording and utilizing digital equipment, a great deal of exceptional production has been done with a basic analog studio setup and some ingenuity, such as the production tricks you'll learn in this chapter.

12.2 PRODUCTION STUDIO SETUP

Before beginning any production work, set up the production studio. This means cleaning up any mess left behind from previous work in the studio. Theoretically, there shouldn't be anything to clean up since good production practice dictates that each person cleans up the production facility after each work session; but theory doesn't always translate into practice, so, if necessary, put away background music CDs, take old tapes off the recorders, zero counters, and clear some working space for yourself. If you need to dispose of a "dull" razor blade, be sure to put a piece of masking or cellophane tape over the sharp edge of the blade before you toss it into the waste basket. (Putting the tape just a bit beyond the blade edge will keep whoever empties the trash from possibly cutting themselves.)

Set all audio console controls to a neutral position. In other words, turn on only those pieces of equipment that you're going to use. Not only is it easier to keep track of and manipulate the volumes for just the equipment you need, but it also prevents any additional audio noise from being introduced into your production work from a piece of equipment that you're not even using.

Clean all tape-recorder heads. If editing has been done recently in the studio, some residual grease pencil may have been left on the heads, and even normal use leaves some oxide material on or near the heads.

Bulk erase the tapes that you plan to use with a degausser. In general, you don't need to worry about erasing anything that may already be on these tapes because the custom in most studios is that any tape left in the production studio is up for grabs. If you want to save something that you've recorded, take the tape out of the production studio.

Take a look at the audio tape, too. If it's excessively worn or damaged in any way, throw it out. If you're using reel-to-reel tape, spool off a few feet from the front of the reel, and throw it away. That part of the tape gets the wear and tear of threading and handling and will wear out before the rest of the tape. If you're producing something that's very important or something that you know will be aired over and over, start with fresh tape.

If possible, use a tone generator to set levels to the tape recorders. (Remember, some audio consoles have a built-in tone generator for this purpose.) At the very least, read some copy into the mic and set a good level on the audio board, and then balance that level with the input levels of the recorders. Do the same for the CD player and other sound sources you'll be using.

This few minutes that you take to set up the studio will make the production go more smoothly. It will also save you time in the overall production process.

As mentioned in the previous chapter, track sheets are used to keep notes regarding multitrack productions. It's also good practice to keep notes while you're doing any production work. For example, if you're using any signal-processing equipment, note the settings for the effect you're producing so that you'll be able to recreate it easily at another time. Also, write down the production music you use for music beds. That way you'll avoid using the same piece of music over and over. Make note of any sound effects or special recording techniques, so you can duplicate what you've produced at some future time, if need be.

When finished production work goes onto audio cartridges, make sure the cart has been erased and cued past the splice. After erasing and locating the splice, listen to the cart for a few seconds. Not only does this confirm that it is properly erased, but playing it helps prevent loose tape problems. (Take the cart out of the program mode on the audio board before recording onto it, or you'll get unwanted echo.)

Always listen to your finished production at least once over the studio monitor speakers, then once over the cue speaker. By listening in cue, assuming the cue signal is in mono, you can tell if a stereo cartridge is out of phase because you'll be barely able to hear the out-of-phase part of the signal. Also listen for copy errors; make sure you said what the copy said. Of course, listen for overall quality. If the production didn't come out the way you think it should have, take the time to redo it.

12.3 TAPE LOOPS

Sometimes you may have a very short production element, such as a sound effect or music bed segment, that only lasts about 10 seconds but needs to be played throughout the length of a spot, say a 60-second commercial. You could record it on an audio-tape cartridge several times back-to-back using a cart editing technique (review section 6.14) or you could just record it on reel-to-reel over and over. A better and quicker technique might be to record a **tape loop**.

As the name suggests, this is a piece of reel-to-reel audio tape that is just looped and spliced together. To produce a tape loop, record the element (let's say a sound effect) on reel-to-reel. With a grease pencil, mark the beginning of the sound and the end of the sound on the back of the audio tape. Cut the tape, using the splicing block diagonal cut, at the beginning sound. Now, so that you don't get confused when you loop the tape and thread it, draw an arrow on the back of the tape in the correct direction of the tape travel. Cut the tape at the end sound, bring the two cut ends together, and splice them into a loop, making sure there is no twist in the tape.

The loop is then threaded onto the recorder by putting two empty reels on the supply and take-up sides and otherwise threading the tape properly, as shown in Figure 12.1. Make sure the tape is threaded in the right direction by checking that the arrow you marked on it is moving past the heads in a left-to-right direction. You may have to provide some tension to the tape (use the smooth surface of the grease pencil) to put any idler arms into "play" mode to make the loop work, or put the recorder into "edit" mode to allow it to play with the idler arm in the off position. Put the reel-to-reel into play, and the tape should now play normally on the recorder over and over until you stop it.

Of course, there is a minimum length of loop that you can create and still thread on most recorders, and an exceptionally long loop will be difficult to handle and require you to string it all over the production studio. You also have to be aware that any sound repeated over and over will eventually be noticeable to the listener, but it will usually work as a background element for 30 or 60 seconds and

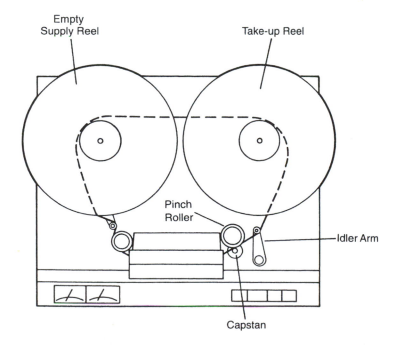

FIGURE 12.1 Threading of recorder for tape loop.

may be just the thing for your particular production situation.

In the digital domain, tape loops are much easier to construct. With some systems, you simply record what you want once, and then use the repeat command to lay it down as many times as you want. Other systems allow you to mark a start and end point on the audio and simply play the defined loop over and over.

12.4 VOCAL STRIPPING

Many creative radio production spots (both parody and straight commercials) have been based on the rewriting of a current popular song. Adding clever lyrics is half the job, but how do you get the instrumental music bed to sing over? Karaoke-style music beds are available for many popular songs, but often they're not cleared for broadcast use, so you must be careful of copyright violations. Some broadcast "mixes" of songs are available that just provide the instrumental backing of a song, and these can be readily used. Broadcasters have also used two main methods of coming up with music beds by **vocal stripping** popular songs. Many 1960s songs were recorded with essentially all the instrumentals on one channel and all the vocals on the other. (Check out some of the early Beatles and Beach Boys songs.) By manipulating the balance control or just rerecording the channel with the music on it, you can eliminate a lot of the vocal. Of course, if some instrumentation is on the vocal channel, you lose that, too, but since this is just a background bed, it may not be very noticeable.

More modern music is often recorded in such a way that the instruments are recorded on either the left or right channel, and the vocals are recorded on both to "appear" to be centered. If you can play back such music so that one channel is out-of-phase with the other, the centered vocals will cancel out and be eliminated or greatly diminished. To strip the vocal, send one channel directly to a mixer. Send the other channel through a device to invert the phase before sending it to another channel of the mixer. Some equalizers will do this, or you may have to employ the talents of your station engineer to accomplish it. Set both mixer channels to equal volume to start, and play the song. You may find that the vocal is greatly reduced, or you may find it necessary to increase or decrease the volume of the inverted signal. You probably will never entirely remove the vocal, but you can reduce it on many recordings. You'll also find that some (or a lot) of the bass is reduced. You can boost the lower end a bit by using some equalization, and you'll want to add some reverb to your "new" vocal to match the music track.

Regardless of the method, not every song will work out well, and, as with most production tricks, you'll find it necessary to do some experimentation to get just the effect you are looking for.

12.5 BACKWARD PLAYBACK

If you can think of a use for a backward sound in your production spot, this trick will be useful. First record the copy in a regular manner on reel-to-reel. Now, don't rewind the tape, but rethread it as shown in Figure 12.2. You may have

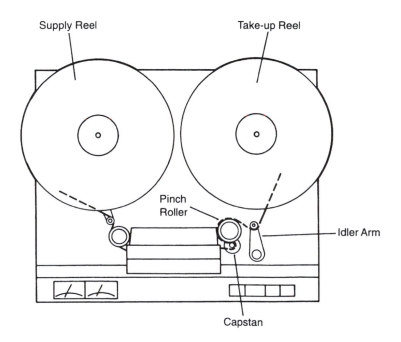

FIGURE 12.2 Threading of recorder for backward playback.

to experiment a little if your recorder's capstan and pinch roller configuration is different than the one shown. In fact, this trick may not work on all recorders. Set up the tape recorder to playback, and when you press "play," you should see that the tape goes through the recorder backward. In other words, the tape goes from the take-up reel to the feed reel, and it plays backward from the end of the spot to the beginning.

Again, with digital equipment, this is much simpler. Most computer audio editing programs have a command that plays material backward.

12.6 REVERSE ECHO

Another production trick used to "punctuate" a key word in production copy is to **reverse echo** the word. Follow these steps to accomplish this effect: First, record the important phrase on reel-to-reel tape recorder #1 (for example, "This gigantic sale saves you . . . fifty percent!"). Be sure to pause before, and really emphasize, the key word(s) (in this case, "fifty percent"). Record it several times at both $7\frac{1}{2}$ and 15 IPS, and leave a lot of room between each "take." Don't rewind the tape, but dub the tape you just recorded to reel-to-reel tape recorder #2 *backward.* Either thread it for backward playback, as previously mentioned, or literally remove the reels and reverse them (i.e., take-up reel becomes feed reel and vice versa). As you're dubbing, you need to add some "echo" to the key word. You do this by having recorder #2's output selectors set to "reproduce" or "tape" and having this recorder in "program" on the audio console. As you increase the gain for this recorder on the console, you are increasing the "echo" effect. You'll have to experiment to find the right amount. Finally, reverse the tape on recorder #2 (this is the backward dub), and play it back. You should hear a nice "sneak," or "pre-echo," of the key word before you actually hear the emphasized word.

Splice this phrase into the copy at the appropriate point, add background music and other effects, and you should have a production spot that stands out from the rest. You'll find that this effect is similar to the slap-back echo mentioned in the previous chapter.

12.7 CAPSTAN WRAPPING

If your production facility has a variable speed tape recorder, and many do, you may not need this technique to produce off-speed effects, but **capstan wrapping** can be useful to vary the speed of any tape recorder. As the name suggests, the procedure is merely to wrap a piece of tape (cellophane tape, masking tape, or duct tape) neatly around the metal capstan of the recorder. Make sure the tape is

wrapped around opposite to the direction the capstan rotates so that the end of the tape doesn't unravel as it rotates. Wrap the capstan either before or after recording. If you record normal and playback with the capstan wrapped, the audio will sound slow; if you record with the capstan wrapped and playback normal, the audio will sound fast. As with most effects, a degree of experimentation is necessary to get just the sound you're looking for. Combining capstan wrapping with a recorder's speed selector and pitch control should allow you to produce a variety of effects. Off-speed recording is often used to create a "chipmunk" voice, and with a little practice you'll get one that is "realistic" and easy to understand. When you are done, make sure you remove all tape and tape residue from the capstan, leaving the recorder in normal operating condition for the next production project.

12.8 TELEPHONE TONES

Even though it's not the best copy writing practice to put phone numbers in a radio spot, local production spots often require including one. You can help emphasize the telephone number by underscoring the reading of the number with "touch tones." You obviously need a touch tone phone and the ability to record the phone. Your production studio might have a phone coupler (review section 8.15) to bring the phone audio into a channel of the audio console. With a little practice, you should be able to punch out the phone number at the exact same time as you say the numbers, recording both on a reel-to-reel. It's a nice way to add an ear-catching "effect" underneath a phone number.

12.9 CONCLUSIONS

Many of the tricks and techniques described here can be accomplished more quickly and more easily with digital audio workstations than with conventional audio equipment. For example, a loop can be made with a few mouse clicks and a "copy" command; the speed of an audio segment (or one word or one syllable) can be varied without wrapping tape around anything. In fact, there's hardly ever anything to clean up around a digital audio workstation because the sounds are stored in the computer; when a new person comes in to start a new production, he or she simply creates a new file on the hard disc.

However, digital audio workstations won't always be at your disposal, and many effects, in addition to the ones discussed in this chapter, can be created with older equipment. The best way to create new effects is by production studio experimentation. If you create something of your own, it will be unique to your station and your production work.

Self-Study

1. The production studio should be cleaned up by _____.
   a) each production person when he or she finishes a session
   b) the first person to use the facility in the morning
   c) the station receptionist
   d) the last person to use the facility at night

2. When you're working in the production studio, you should _____.
   a) turn on all the equipment even if you don't plan to use it
   b) use only fresh tape
   c) set levels using the tone generator
   d) not keep notes because it causes you to do the same style of production over and over

3. To check a finished production spot for out-of-phase problems, listen to it on _____.
   a) stereo monitor speakers
   b) stereo headphones
   c) stereo cue speakers
   d) a mono cue speaker

4. A production trick that plays back music with one channel out-of-phase with the other is trying to _____.
   a) create reverse echo
   b) record in mono
   c) build a tape loop
   d) strip the vocal

5. Most modern reel-to-reel recorders can be threaded in a way that will make them play backward without having to reverse the supply and take-up reels.
   a) true
   b) false

6. If the capstan of a reel-to-reel recorder is wrapped with tape before recording, and then the tape is removed before playback, the audio will sound _____ when you play the tape.
   a) slow
   b) fast
   c) echoed
   d) normal

7. A production technique that allows you to repeat a piece of audio over and over is _____.
   a) vocal stripping
   b) backward playback
   c) tape looping
   d) capstan wrapping

8. Which of the following are *not* correctly matched?
   a) degausser—erase audio tapes
   b) tone generator—create telephone "touch tones"
   c) track sheet—keep production notes
   d) reel-to-reel recorder—create "tape loops"

9. Which production trick would most likely be utilized to create a "chipmunk" voice?
   a) backward playback
   b) capstan wrapping

    c) reverse echo

    d) vocal stripping

10. As a final review of production tricks, match the techniques and equipment in the top list (1, 2, 3 . . .) with the most logical reason for which they would likely be used in the bottom list (o, s, a . . .), and then select the correct set of answers from the sequences shown in a, b, c, or d below.

    1. _____ creating a tape loop

    2. _____ using a tone generator

    3. _____ backward recording

    4. _____ telephone tones

    5. _____ degaussing

    6. _____ vocal stripping

    7. _____ reverse echo

    8. _____ capstan wrapping

    o. having bird chirps be heard over and over

    s. adding touch tone sounds to a commercial

    a. slowing down a portion of audio

    j. hearing a jumbled sound that occurs when sound recording goes in reverse

    l. adding new lyrics to an old song

    w. emphasizing a particular word

    c. calibrating volume levels between an audio console and a recorder

    e. erasing an audio tape

    a) 1.j 2.c 3.o 4.s 5.e 6.l 7.w 8.a

    b) 1.j 2.s 3.o 4.c 5.e 6.l 7.w 8.a

    c) 1.o 2.e 3.j 4.l 5.c 6.s 7.a 8.w

    d) 1.o 2.c 3.j 4.s 5.e 6.l 7.w 8.a

■ ANSWERS

If You Answered A:

1a. Right. All production people are responsible for their own cleanup.

2a. No. This can introduce unnecessary noise. (Reread 12.2.)

3a. No. This would not let you hear a phase problem. (Reread 12.2.)

4a. No. Reverse echo is a different production trick. (Reread 12.4 and 12.6.)

5a. Yes. This is the correct response.

6a. Wrong. If you record normal and playback with the capstan wrapped, the audio will sound slow. (Reread 12.7.)

7a. No. This is another production technique. (Reread 12.3 and 12.4.)

8a. Wrong. A degausser is used to erase audio tapes. (Reread 12.2.)

9a. No. While you might get a sound similar to a chipmunk, it would be backward, and you wouldn't be able to understand it. (Reread 12.5 and 12.7.)

10a. No. You're confusing loops and backward playback. (Reread 12.3 and 12.5.)

If You Answered B:

1b. Wrong. No one would want to be first. (Reread 12.2.)

2b. No. It's fine to reuse audio tapes if they're not excessively worn. (Reread 12.2.)

3b. No. This would not let you hear a phase problem. (Reread 12.2.)

4b. No. You may be getting this confused with listening to production work in mono to detect phase problems. (Reread 12.2 and 12.4.)

5b. No. Most recorders can be threaded to play backward. (Reread 12.5 to learn how.)

6b. Yes. The technique described will speed up the audio.

7b. No. This is another production technique. (Reread 12.3 and 12.5.)

8b. Correct. A tone generator is used to set levels for recording.

9b. Yes. Off-speed recording can be made by a capstan wrapping technique.

10b. Wrong. You're confusing telephone tones and a tone generator and also tape loops and backward playback. (Reread 12.2, 12.3, 12.5, and 12.8.)

If You Answered C:
1c. Naturally not. (Reread 12.2.)
2c. Yes. Setting levels with the tone generator is a good practice.
3c. No. This won't let you hear phase problems. (Reread 12.2.)
4c. No. This is a different production trick. (Reread 12.3 and 12.4.)
6c. No. You may be getting this confused with the reverse echo technique. (Reread 12.6 and 12.7.)
7c. Yes. This is the correct answer.
8c. Wrong. A track sheet is used to keep notes during multitrack recording. (Reread 12.2.)
9c. No. This is a way to punctuate a key word. (Reread 12.6 and 12.7.)
10c. You made many mistakes. (Reread the entire chapter.)

If You Answered D:
1d. No. This wouldn't be fair to whoever was last. (Reread 12.2.)
2d. No. It's good practice to keep notes because you may have to recreate a particular production spot later. (Reread 12.2.)
3d. Correct. Listening in mono (through any speakers or headphones) will allow you to hear phase problems.
4d. Correct. This is one technique for stripping vocals.
6d. No. If the capstan size was changed between recording and playback, it won't sound normal. (Reread 12.7.)
7d. No. This is another production technique. (Reread 12.3 and 12.7.)
8d. Wrong. A reel-to-reel recorder is used in creating a tape loop. (Reread 12.2 and 12.3.)
9d. No. This is a trick to eliminate vocals from songs. (Reread 12.4 and 12.7.)
10d. Correct. You have now finished this chapter.

Projects

■ PROJECT 1

Create a tape loop.

Purpose

To enable you to develop skills utilizing this production trick.

Advice, Cautions, and Background

1. Choose a sound effect or music segment that realistically might have to be repeated, such as a telephone ring, rather than just a random bit of audio.
2. Review section 12.3 before beginning this project.
3. If you have access to digital equipment, you can use it to create the same effect rather than follow the instructions given here.

How to Do the Project

1. Select a sound effect that's approximately 10 seconds in length.
2. Record the effect onto reel-to-reel tape.
3. Mark and cut the beginning and end points of the effect on the tape, and splice the two ends together. Remember to mark the correct direction of travel of the tape so that you can thread the loop properly.
4. Thread the loop on the reel-to-reel recorder and play it. Remember to put the recorder into edit mode, or hold the idler arm in the play position.

5. If the loop works as you expected, record 60 seconds of it onto a cassette. If it sounds awkward or just didn't come out as expected, redo it.
6. Select a segment of music that's 30 seconds in length.
7. Repeat the above steps to create another tape loop.
8. Record about 2 minutes of your music loop onto the cassette. You may have to get creative to make the longer loop play correctly on the recorder.
9. Put the cassette tape and your tape loops into an envelope. Put your name and Tape Loop Project on the envelope and the cassette tape.
10. Turn in the completed materials to the instructor to receive credit for this project.

■ PROJECT 2

Create a reverse echo effect.

Purpose

To enable you to develop skills utilizing this production trick.

Advice, Cautions, and Background

1. Review section 12.6 before beginning this project.
2. Although this project could be configured other ways, it's assumed that you have two reel-to-reel recorders in your production studio.

How to Do the Project

1. On recorder #1, record the phrase, "These low, low prices are good only until . . . Friday." Record it several times. Be sure to pause just before the key word ("Friday"), and emphasize this word. Leave some blank tape (space) between each recording.
2. Dub this tape onto another recorder *backward*. Either thread recorder #1 for backward playback, or actually reverse the feed and take-up reels to make it play backward.
3. As you are recording on recorder #2, you'll need to add some "echo" on the key word. You can do this by having recorder #2's output switches set to "reproduce" or "tape" and having the recorder in program on the audio console. By increasing the gain on recorder #2, you'll increase the "echo" effect.
4. You may have to experiment with this several times to get the effect just right.
5. Reverse the tape on recorder #2 so that you can play it back. As it's played back, you should hear a "pre-echo" of the key word ("Friday") just before you hear the actual emphasized word.
6. Label the finished tape with your name and Reverse Echo Project, and turn it in to the instructor to receive credit for this project.

APPENDIX A

Production Situations

A.1 INTRODUCTION

Some additional production tips that apply to particular types of radio programming—namely, music, commercials, news, public affairs programs, call-in talk shows, play-by-play sports, and drama—are discussed in the next few pages. Entire books have been written about each of these types of programming, so you won't be given the in-depth knowledge you need to perfect any of the program forms. This appendix will, however, get you started in the right direction. Experience and advanced training can then propel you into more specialized skills. Although music, commercials, news, public affairs programs, call-in talk shows, play-by-play sports, and drama are major forms of radio programming, other forms do exist. For example, some stations have children's programs, and others occasionally produce documentaries. Still others are no doubt open to new forms of programming that may become popular in the future.

A.2 MUSIC

Music constitutes the largest percent of radio station programming and is usually introduced and coordinated by a **disc jockey**. If you become a disc jockey, you'll be spending most of your time in the on-air studio doing all your production work live. This is actually a more challenging production situation than that in the production studio, where you have the time and luxury of rerecording until you get things just the way you want. On-air broadcasting is fast-paced, pressure packed, and, for most people, fun.

Although the main element of the programming is music, the main duty of the disc jockey is talking. Much of this talk involves introducing music. For this, your announcing style must fit the format of the radio station. For example, fast-paced, high-energy, rapid-fire speech is not appropriate for a classical music or big band format but may be required at a contemporary hit radio station.

Have a variety of ways of getting into and out of the music. Many beginning announcers latch onto one record introduction and use it over and over ("Here's a classic from The Beatles . . . "; "Here's a classic from Bob Dylan . . . "; "Here's a classic from . . . "). If you have trouble thinking of clever material, read the liner notes on the CD or record.

This will often give you an idea for something to say that's unusual or informative.

Your station may have certain policies regarding what you say and how you say it. For example, some stations require you to talk over the beginning and ending of records to prevent people from recording whole songs off the air. Other stations have particular slogans ("All hit radio") that they want repeated at regular intervals.

As a disc jockey, you'll also be talking about things other than music. For example, you may need to give the time, temperature, commercials, weather, news, or traffic reports. Or you may introduce other people, such as the newscaster, who will give some of this information. Station policy will probably dictate whether you must be very formal about these introductions or have the latitude to banter with the other person. At some stations, the good-natured joking between, say, disc jockey and helicopter traffic reporter is part of what keeps listeners tuned in.

Regardless of what's going on within the live production situation, always assume the mic may be open. Don't say anything that you wouldn't want to go out over the air. This includes personal conversations and, of course, indecent language. Many studios have an on-air light inside the studio as well as the one outside the door. This inside light is to alert the announcer to the fact that the mic is live, but the best rule is to assume the mic is always on.

Obviously, as a disc jockey, you need to be proficient at operating the equipment so that you can cue up and play tapes and CDs. But this involves more than just equipment manipulation since you're normally operating equipment and talking. This means you have to plan ahead. For example, if you have to join a network, you must know exactly when and how to do it. If you aren't sure whether you've cued up a tape, redo it. You must have everything cued up and ready to go, and have a routine for what you are doing. It's also good practice to have an alternative if something goes wrong. If the cartridge you want to play doesn't fire, have in mind what you'll do immediately. A good announcer can overcome most miscues so that the listening audience doesn't even know that anything went wrong.

Make sure you have previewed the record, especially to know how it begins and ends. This will insure that you avoid **walking over** (beginning to outro a song before it's really over) a false ending. Previewing also helps you know how much instrumentation there is at the beginning of the record before the vocal starts. In this way you can talk over the instrumental but not the vocal part of the beginning. Sometimes timing information is provided on the CD or cartridge label, and this helps you with **voice-overs** on the intro and outro of the song. In any case, there is no excuse for a disc jockey playing a CD on the air that he or she isn't completely familiar with.

Plan how you'll get from one piece of music to the next if you play them consecutively. You might want to review the sound transitions mentioned in Chapter Two. Remember to work for a variety of ways rather than using the same method time after time. Listen to the on-air monitor frequently if not continuously. Most audio consoles allow the announcer to hear the program line, the audition line, or the on-air signal; however, only the on-air signal allows you to hear exactly what the listener hears.

When your shift is over, clean up around the studio. It's also thoughtful to pull the first couple of songs or the first commercial break material for the next announcer so that he or she can get off to a smooth start. Playing music on radio is a hectic but rewarding job. Disc jockeys are, in every sense of the word, production people. Not only do they have to manipulate all the broadcast equipment, but they must add the element of announcing and present it all live and within the fast pacing of the radio station's format.

A.3 COMMERCIALS

Most radio production people are involved with producing commercials. At some stations it's a full-time position, and at many stations the disc jockey spends a portion of their workday doing commercial production work. At their simplest level, commercials are simply copy that the announcer reads over the air. At their most complex, they are highly produced vignettes that include several voices, sound effects, and music. In between are testimonials by famous people, an announcer followed by a cartridge, dialogue between two actors, and an announcer over music.

Commercials are usually exactly 30 seconds or exactly 60 seconds. Because they are inserted within other programming elements, they must have exact times. Otherwise part of the commercial will be cut off, or there will be dead air. At some stations commercials are brought on and taken off by computers, which are very unsympathetic to anything that's too long or too short.

Anyone reading commercial copy should use a natural, sincere style. Reading commercials in a condescending manner is definitely uncalled for. Commercials pay station salaries and should be treated with respect. If you're to read commercial copy live, you should read it over ahead of time to minimize stumbling on words. If the commercial involves both reading live and playing a cartridge, make sure you rehearse the transition between the two. Sometimes a cart sounds like it's ending, but it actually has additional information—perhaps new store hours or a bargain price. If you're to read material after the cart ends, make sure you don't read over the top of the carted information.

Probably the most basic form of commercial you'll produce involves an announcer over music. The usual format for this is as follows:

a. music bed at full volume for a few seconds
b. music bed fades under and holds
c. voice-over read on top of music bed
d. music bed brought up to full volume at end of voice-over for a few seconds
e. music bed fades out

Although this seems simple enough, it won't come easily until you've practiced and accomplished it many times. Not only must you be concerned with timing, but you have to determine how much music to use to establish the spot, balance the levels between voice and music bed, select appropriate music, and correctly manipulate the broadcast equipment. Usually you do all of this at the same time, although it is possible to record the voice-over and mix it with music at a later time.

Highly produced spots usually take a relatively long time to prepare. Minidramas, which start with music and involve two people bantering to the accompaniment of sound effects, involve a great deal of preproduction, rehearsing, mixing, and editing. They are, however, among the most challenging, creative products a radio production person handles. Often different parts of them are recorded at different times. The talent will record their parts and then leave, especially if they're highly paid talent. Then the production person will mix the talent skit with music and sound effects.

As you can see, mixing complex commercial spots (or any type of production) can be complicated. If you miscue something or have the balance between two elements set up wrong, you have to redo the entire spot. Production work of this type can more easily be accomplished by using a multitrack recorder.

A.4 NEWS

If you're involved with radio production, you're likely to do some news work at some point in your life. In some instances the disc jockey may just **rip and read** from the news service on the hour or half hour. In a worst case scenario, the announcer would literally just rip some news copy off the wire and read it on-the-air. As we'll note in a moment, this is *not* good broadcast practice. At the other extreme are all-news stations, which usually have separate on-air people for news, weather, sports, traffic, commentaries, business news, and the like. This is a complicated format that necessitates using people who are very knowledgeable in their specialties and who have established themselves and their voices as personalities. Whatever situation you find yourself in, some knowledge about newscasting will prove useful.

Even if you're expected to just rip and read, you should give some time and thought to your news presentation.

First, you must decide which news to present. Of course, you probably have a station news format to follow, but you should also try to select those items that are most likely to be of interest to your listeners. You may find that you need to do some rewriting of the wire service news and write some transitions to take the listener from one story to the next.

Timing is important on a news break. Make sure you read your news copy in the time prescribed. Sometimes beginning newscasters run out of news to read before the newscast time is up. To prevent this, you should pad your newscast with some extra stories that you can cut if you have to but that provide a cushion if you need extra material. You'll usually need to get commercials in at an appropriate time during a newscast, so make sure you know when this happens and are prepared for it.

Don't ever read a newscast cold. Read it over first so that you're familiar with the material. Rewrite anything that isn't natural for you, such as tongue-twisting phrases that you might trip over. Also rewrite if there are long sentences that make you lose your breath before you finish them. Avoid too many numbers or facts jammed into a single sentence. Whenever something is unclear, rewrite to make it simple and easy to understand. Remember, broadcast news should be conversational and written for the ear, not the eye.

When a news story includes the actual voice of the person in the news, such as the mayor commenting on the new city budget, that segment is called an **actuality**. Most radio news operations strive to include many actualities within a newscast because these bring life to the news. It's more interesting to hear the mayor's comments than the voice of an announcer telling what the mayor said. Actualities are best reserved for opinion or reaction. The facts about the story can be given by the reporter, and then the newsmaker can say what he or she thinks about the subject.

Obviously, many news actualities are gathered in the field with portable tape recorders; you can, however, also make use of the phone to gather them. Many small radio stations have one-person news departments, and in these cases the phone actuality is especially crucial. In some situations, newsroom phones are semipermanently hooked up to tape recorders. But if not, you can use **alligator clips** to tap into the phone signal. Remove the earpiece cover on the phone receiver, attach the clips to the two prongs within the earpiece, and send the signal to the input of your recorder. Another way to record from the phone is to use a contact mic that's specifically made to attach to the phone for this purpose. You must obtain permission to record someone for broadcast use, and your station newsroom probably has specific guidelines to follow for doing this.

Actualities generally need to be edited. For this, you should use all the editing techniques presented in Chapter Six, and then add one very important rule: Make sure that when you edit, you don't change the meaning of what someone has said. Ethical news procedures dictate that a great deal of care be taken in this area because the elimina-

tion of a single word can significantly alter a news report. For example, "The mayor did not agree with the city council's decision. . . . " could easily become just the opposite if the word "not" were edited out, leaving, "The mayor did agree with the city council's decision. . . . " When you're editing and need to eliminate part of what a person said, either because it's too long or it's irrelevant, try to match voice expressions where the statement leaves off and where a new one begins. Edits are usually best if made at the end of thoughts because a person's voice drops into a concluding mode at that point. Be careful not to edit out all the breaths the person takes because this will destroy the natural rhythm. People do breathe, and actualities without any breathing sound unnatural. If an edit is going to be too tight because the person runs words together, add in a little background noise, or add a breath from a different place on the tape. Either of these will make for a natural-sounding pace between two edit points.

Try to maintain a constant background level throughout the actuality. To do this, you may have to mix in background noise from one part of the tape to another part. If you mix narration with an actuality, make sure you maintain the same level for both, and try to have background noises that are similar—or at least not jarring.

When splicing, cut loosely if in doubt of where you actually want to make the edits. It's much easier to take more out later than it is to put material back in once it has been edited out.

Whether you're a disc jockey reading a few newscasts or a broadcast journalist at an all-news station, you may wish to consult one of several good books available on techniques and ethics of gathering, writing, editing, and presenting news. Some of these books are listed in the Suggested Reading section at the end of this book.

A.5 PUBLIC AFFAIRS PROGRAMS

Public affairs programming usually consists of long programs that explore a news or community issue in depth. Some public affairs programs take a lighter tone and profile an individual or group. The typical public affairs show is a half-hour interview or discussion between host and guest(s). Although some stations have a specific public affairs host, many stations delegate this responsibility to a news person or an announcer.

The key to good public affairs programming is proper preparation. The host needs to research the subject and the guest before doing the program. Not only will this provide background, but it should enable the host to make a list of questions to ask.

Asking the right questions really means asking good questions. For example, ask questions that require more than a simple yes or no answer. Don't ask, "Do you agree with the mayor's new policy regarding the police?" but rather, "What do you think of the mayor's new policy regarding the police?" Ask short, simple, and direct questions. The question, "Given the salaries of employees and the possible raises they will receive, what do you think the effects will be on the social security system and the GNP?" will most likely get a response of, "Huh?" Break complex questions down into a number of questions such as, "How do you think increasing salaries five percent will affect the GNP?" Ask questions that don't require long answers. Don't ask, "What would you do to improve the city?" but, "What is the first thing you would do to improve the city?"

Asking good questions also means knowing how to handle the answers you get. For example, if the answer is too wordy, ask the person to summarize the response. If the answer is muddy or unclear, ask the material over again in smaller parts. If the answer is evasive, come back to it later, or ask it again from a different angle. If the response gets off track, redirect. And if the response goes on and on, interrupt politely and redirect.

Listen carefully to what the guest says so that you can ask appropriate follow-up questions. Sometimes interviewers become so engrossed in thinking about the next question that they miss an important point the guest has made that could lead to something significant. Although you should write questions ahead of time so that you remember to cover all important points, don't slavishly stick to those questions. In all probability, when you ask question one, the guest will also answer questions three and seven, so you must constantly flow with the conversation. The more you can lead off what the guest says, the more natural the interview will appear. But make sure you do get the information you want.

If you're going to be recording the program at a remote location, make sure all the equipment works before you leave the station. This also holds true for recording in the studio; check everything out before your guest arrives.

You may be using a single microphone, so talk at the same level as the interviewee. As a broadcast professional, your voice may be stronger than the interviewee's, in which case lower your voice or else keep the mic farther away from yourself than from the guest. If you're talking with several guests, move the mic back and forth as each person talks. Also, follow speakers if they move their heads. Never let the guest take the microphone. He or she probably won't know how to use it properly, and you'll lose control of the conversation.

When you have a number of different guests, identify them frequently because the listener has difficulty keeping track of the various voices. Even a single guest should be re-introduced during a 30-minute program. Not only does this remind listeners who your guest is, but it introduces him or her to those listeners who joined the program in progress.

When you record at a remote location, listen to at least part of the tape before you leave to make sure you have actually recorded something. It's embarrassing to get back to the station and discover that you have nothing on the tape, and it can be impossible to rearrange the interview for another time. If you're doing the interview in the studio, make sure you have a recording before the guest leaves.

Many public affairs programs are aired live or "as taped," but sometimes it's necessary to edit the tape. If you're going to be editing extensively, make a log of the material so that you know what is on the tape and where it is.

An understanding of the public affairs programming concepts and techniques, especially interviewing, will help you handle many types of production situations.

A.6 CALL-IN TALK SHOWS

Not everyone is cut out to handle hosting a call-in talk show. You must be fast on your feet and able to ad-lib in an entertaining and effective manner. For many programs, you're expected to have more than broadcast production knowledge. For example, a sports program or radio psychology show requires a host with some expertise in those areas. Usually the telephone talk show host has an engineer handling the equipment so that the host can concentrate on dealing with the callers, but this is not always the case, especially in smaller market radio.

A telephone talk show host has to be able to handle people tactfully (or, in some cases, abrasively, if that's the style of the program). As when interviewing for public affairs programs, the host must remain in charge of the program. Many of the other principles of public affairs interviewing also apply to call-in shows.

If you have a guest that people are asking questions of, you should give information about the guest and redirect questions if they're not understandable. The host must be able to establish good communication, without the aid of body language.

A.7 PLAY-BY-PLAY SPORTS

Many announcers also handle some sports broadcasting duties, and **play-by-play** (**PBP**) announcing skills can be a valuable asset for anyone entering the radio business. As a radio sports announcer, you must keep up the chatter, and you must describe completely what is happening. Since the listeners don't have the video image of TV, you have to be their eyes. Sports announcers often work in a team, with one announcer providing the play-by-play while another offers color commentary along with game statistics.

The PBP sportscaster operates remote equipment designed for sports broadcasting. Most of it is similar to studio equipment but extremely portable. A small audio console and headset microphone make up the bulk of sports remote equipment. The headset mic arrangement allows the sportscaster the use of both hands and keeps the counter or tabletop free for equipment and stat sheets.

The signal is sent from a remote console to the station by phone line or a portable transmitter that relays the signal to a special receiver at the station. You should also have some other way of communicating with the studio, such as a separate telephone line. This way you can talk with the station about production matters that should not be heard by the audience. For example, commercials are usually played from carts at the station, and you might need to coordinate when and how many will be played during a break in the action. Many remote consoles have some type of talkback feature that allows off-air communication with the radio station.

Pregame preparation is very important for sports broadcasting. Gather all the information and facts about the teams and players, make sure all the equipment is ready to go, and get to the game in plenty of time to check out the broadcast booth, so you know everything is in order.

Sportscasting is one of the most glamorous aspects of radio production, but it is time-consuming and requires quick thinking plus a thorough knowledge of the sport.

A.8 DRAMA

Drama isn't produced very often on radio anymore, but when it is undertaken, it employs the use of many production skills, such as microphone techniques and the blending of sound effects and music into the production. The setting and action must be conveyed by the talent, sound effects, and music. Drama is best recorded with one microphone, but you may need to have people with louder voices position themselves a little farther away from it. It's also important to keep **perspective** during a production. In other words, people who are coming toward the supposed location of the play should sound far away at first and then nearer. The best way to do this is for the talent to walk away or toward the mic. You can mark the studio floor as to where they should start or stop talking to be on mic or off mic.

Sometimes one actor performs several parts. The actor needs to be able to create distinctly different voices in order to do this. In fact, all voices used within a drama should be distinctively different because audience members have trouble distinguishing one character from another. Talent should avoid using an affected radio voice but rather talk in their character's natural, conversational style.

Sound effects can be simulated in the studio, usually on a different microphone than the talent is on. Or they can be prerecorded or taken from a sound effects CD and edited into the production at a later time. They can be used to establish locale, tell time, create mood, indicate entries and exits, and establish transitions. Music is also an effective way to create mood and establish transitions. Most radio dramas open and close with music that's appropriate to the overall theme of the drama.

Additional Production Projects

■ PROJECT 1

Record a 15-minute rock music, classical music, or country music disc jockey show.

Purpose

To enable you to develop disc jockey skills.

Advice, Cautions, and Background

1. You may do as many of these formats as you wish, and get credit for each. In other words, you could do both a rock show and a classical show, but don't do two of the same type.
2. You'll need different voice intonations for each style. For example, your talk on a rock show might be fast and bubbling, but for a classical show you would be more subdued.
3. Listen to several disc jockeys who do the type of show you are planning to do, and pick up whatever ideas you can, but you shouldn't try to just copy what another disc jockey is doing.
4. It's assumed that you have completed this entire text before you attempt this project.

How to Do the Project

1. Select the music you wish to play, and time it. You'll be playing each piece in its entirety. Except for classical music, selections shouldn't be over five minutes. Try to establish a mood or theme for your 15-minute segment.
2. Plan and write any commercials, station breaks, or other information you wish to include. Think through how you'll introduce the various music selections.
3. Make sure you have a total of approximately 15 minutes of material, including music and talk.
4. Use your own creativity and style. Vary your pitch, volume, and tone so that your voice doesn't become monotonous. Remember that the listener can't see you, so facial expressions and gestures have no effect.
5. Record the program. Make the program exactly 15 minutes. This shouldn't be much of a problem because you can fade out music, back-time, or ad-lib at the end.
6. If it's good enough, give the tape to the instructor. If not, do it again until it's acceptable. Put your name and Disc Jockey Project on the tape.
7. Turn in the completed tape to the instructor to receive credit for this project.

■ PROJECT 2

With several other students, record a 15-minute news program.

Purpose

To further develop news skills.

Advice, Cautions, and Background

1. Begin work on this fairly early. Since it will be a team effort, you'll need time for coordination.
2. For your news team, select reliable people who won't be dropping out of the project. It's annoying to have to change your format after you begin.
3. It's a good idea to select a producer who will handle coordination.
4. Be sure to work up transitions from one type of news to another, as this will probably be the most awkward part of the newscast. Also, plan the beginning and ending.
5. It's assumed that you have completed this entire text before you attempt this project.

How to Do the Project

1. Decide what each person will do. There are many different ways of organizing this, all of which could probably be found at some radio station somewhere. Following are some things to consider:
   a. Decide what program elements you want to include, such as international news, national news, local news, sports, weather, editorial, commercials, human interest.
   b. How do you plan to handle transitions? For example, each person could introduce the next one, or one anchor could introduce all the others.
   c. How do you plan to handle writing and announcing duties? Does each person write and read his or her own news, or do some people write the news and others announce?
   d. How do you plan to handle engineering? One person could be responsible for levels, several people could engineer, or you could preset levels, and no one would be responsible during the actual newscast.
2. Decide on time allocations for the various segments of news so that you total 15 minutes.
3. Check your organization and ideas with the instructor.
4. Set a definite date to do the taping.
5. Have a dry run of the program, and be sure to check timing.
6. Tape the program and listen to it. If it's good enough, turn in the tape to your instructor. If not, redo it.
7. On the tape write NEWSCASTING PROJECT, and list the names of those participating.
8. Give the instructor a separate sheet of paper that lists everyone's name and the duties each fulfilled for the production. Turn in this sheet and the completed tape to receive credit for this project.

■ PROJECT 3

Record a 5-minute interview show in which you are the interviewer.

Purpose

To prepare you for this very common type of broadcasting situation.

Advice, Cautions, and Background

1. Your interview must be exactly 5 minutes. Making an exact time without having an awkward ending will probably be the hardest part of the project, but it's a lesson worth learning because broadcasting is built around time sequences.
2. Don't underprepare. Don't fall into the trap of feeling that you can wing this. In 5 minutes you must come up with the essence of something interesting, and you can't do this unless you have an organization of questions in mind. You'll also only be able to record the interview once. You can't redo this project.
3. Don't overprepare. Don't write out the interview word for word. It will sound stilted and canned if you do.
4. Five minutes is actually a long time; you'll be amazed at how much you can cover in this time.
5. As interviewer, don't talk too much. Remember, the purpose is to get the ideas of your guest over to the audience, not your own ideas.
6. It's assumed that you have completed this entire text before you attempt this project.

How to Do the Project

1. Select someone in class to interview. It's not mandatory to do this. You can interview a friend, but it will probably be easier to do it with someone in class.

2. Decide what the interview will be about. You may select any subject you wish. You could talk about some facet of a person's life or his or her views on some current subject, or you could pretend the interviewee is a famous person.
3. Work up a list of questions. Make more than you think you'll actually need just in case you run short.
4. Think of a structured beginning and ending for the show since those will probably be the most awkward parts.
5. Discuss the interview organization with your guest so that you are in accord as to what is to be discussed.
6. Tape the interview, making sure you stop at 5 minutes. Listen to it, and check that it has recorded before your guest leaves. You are finished with the project once the interview is taped. Even if it didn't come out as you had hoped, do *not* redo this project.
7. On the tape write INTERVIEW PROJECT and your name. Give the interview tape to your instructor to receive credit for this project.

◼ PROJECT 4

Record a 10-minute play-by-play broadcast of some sports event.

Purpose

To allow those interested in doing sports broadcasts to develop skills in this area.

Advice, Cautions, and Background

1. This must be a real situation. Don't just pretend that you're seeing some sports event. You must actually do it from a sports event or else from a TV broadcast of a sports event.
2. Be sure to do your homework. Preparation is important so that you can keep announcing even during lulls in the action. Make sure you know enough about the sport and the players before you attempt the broadcast.
3. Make sure you can pronounce all the players' names correctly.
4. If you go to a game, try to situate yourself in a location where there won't be too much crowd noise to interfere with your recording. If you do the project from a TV set, you will, of course, have to turn down the volume.
5. You may wish to rehearse silently or out loud or by recording several times before you actually do your 10-minute recording.
6. It's assumed that you have completed this entire text before you attempt this project.

How to Do the Project

1. Select the sporting event you wish to use.
2. Attend the event, and take a portable tape recorder; or turn the volume down on your TV set, and get your equipment ready to do a play-by-play.
3. At some point start your broadcast, and continue recording it for about 10 minutes. Try to start at some logical spot, such as the beginning of a quarter, inning, or round.
4. On the tape write your name and PLAY-BY-PLAY BROADCAST PROJECT. Give the completed tape to your instructor to receive credit for this project.

◼ PROJECT 5

With several other students, record a radio drama.

Purpose

To allow you to practice your dramatic abilities and to give you the experience of performing radio dramas.

Advice, Cautions, and Background

1. Make sure you have a group of congenial, dependable people. You don't want to start rehearsal and then have to switch cast members.
2. Sound effects will probably give you the most trouble. Make sure you have all that you need.

3. Participants can get credit for this assignment without being on-air. For example, you might want to have one person in charge of sound effects, one in charge of engineering, and one as director.
4. If you want to write a play rather than use one that's already written, this is fine.
5. One person can play more than one role in radio by changing voices.
6. It's assumed that you have completed this entire text before you attempt this project.

How to Do the Project

1. Select the group and choose a director.
2. Have the director or the group select a play. The following are some books you may be able to find in a library that contain radio plays:
   a. *Radio Workshop Plays*
   b. *Creative Broadcasting*
   c. *Radio's Best Plays*
   d. *Radio Plays For Young People*
3. Decide what each person is going to do for the production.
4. Check with the instructor about what you're planning to do and what problems you expect to experience.
5. Make sure that everyone gets a copy of the script. You can check with the instructor about ways to get the script duplicated.
6. Have set times to rehearse. Rehearse section by section both with and without microphones, and then put the whole thing together. Rehearse music and effects as well as dialogue.
7. When the production is polished enough, record it. Listen to it, and if it's good enough, give the tape to the instructor. If not, redo it.
8. On the tape write RADIO DRAMA PROJECT and the names of all the people who are involved.
9. On a sheet of paper, list the names of all involved, and write what each did for the production. Give this sheet and the completed tape to the instructor to receive credit for this project.

■ PROJECT 6

Record an air check tape.

Purpose

To instruct those interested in doing on-air broadcasting how to make an audition tape, something required when applying for a job.

Advice, Cautions, and Background

1. To apply for on-air jobs in broadcasting, you will send your resume and an air check tape to many stations. An air-check tape is an audio tape of 5 to 10 minutes that you record to show how you handle on-air broadcast situations.
2. Ideally, an air check tape is an edited-down sample of your actual on-air work, but if you aren't on-air on a regular basis, a simulated air check can be put together in the production studio.
3. Try to make the tape as general as possible so that it could be sent to several different types of stations.
4. Put those things you do best at the beginning of the tape. Many potential employers don't have time to listen past the first few minutes and will rule you out if they don't like the beginning. Don't structure the tape so that it builds to a climax because probably no one will listen that far.
5. Feel free to use things you've done for other projects for your tape.
6. Keep the pace of the tape moving. Don't do any one thing for too long.
7. Try to put something unique near the beginning of the tape so that your air check is remembered.
8. It is assumed that you have completed this entire text before you attempt this project.

How to Do the Project

1. Plan what you intend to include in your tape. An air check format might include ad-lib introductions to a few songs (either fade out the music after a few seconds, or edit to the end of the songs so that the listener doesn't

have to hear the whole song), a short newscast, some more song introductions, and a commercial or public service announcement. If you can do play-by-play, you might want to put a short sportscast at the end. There is no standard format, so do whatever showcases your talent best.

2. Plan the order of your tape. Make it sound like a continuous radio show as much as possible.
3. Record your tape and listen to it. Redo it if it doesn't present good broadcast skills.
4. On the tape, write your name and AIR CHECK PROJECT. Turn in the completed tape to the instructor for credit for this project.

■ PROJECT 7

Record a 30-minute original radio program.

Purpose

To give you experience with more complicated production, to sharpen skills learned in other projects, and to utilize your creative ideas.

Advice, Cautions, and Background

1. You're to produce something that can be done on radio but isn't done very often. Therefore, this can't be a regular disc jockey show. It must be an original idea.
2. Start working on this right away because your preproduction time will probably be lengthy.
3. Check with the instructor frequently about your ideas, problems, and progress.
4. Don't get in over your head. Think up an unusual idea, but make sure it's something you can actually complete.
5. Remember, this is to be done so that it could be broadcast, and therefore it must comply with all FCC rules about profanity and indecent language.
6. Make sure the technical quality of the project is good.
7. It's assumed that you have completed this entire text before you attempt this project.

How to Do the Project

1. Think up an idea of some sort of program that could be aired on radio and isn't being done to any great extent. You can pretend the program is part of a series or that it's a single program. Ideas that have been used by students in the past include: a narrative of advice about hitchhiking, a composite of humorous material, and an interview with a 6-year-old.
2. Check your ideas with the instructor.
3. Begin putting the program together in whatever way your particular idea demands.
4. Check with the instructor before you do the final taping and at any other time when you need help or advice.
5. Tape and edit (if necessary) so that you have a finished program that's 30 minutes in length.
6. On the tape write your name, the program title, the exact time of the program, and a very brief summary of the content.
7. Give the completed tape to the instructor for credit for this project.

Glossary

A-B miking   A method of stereo miking where one mic feeds the right channel, and another mic feeds the left channel.

Absorption   The process of sound going into walls, ceilings, and floors of a studio.

Absorption coefficient   The proportion of sound a material can absorb in relation to the sound it will reflect back. A coefficient of 1.00 means that all sound is absorbed in the material.

Acoustic suspension   A speaker enclosure design that consists of a tightly sealed box that prevents rear sounds from disrupting main speaker sounds.

Actuality   A voice report from a person in the news rather than from the reporter.

Adaptive Transform Acoustic Coding   A data compression system used for MiniDiscs.

Adhesion   A condition that occurs when one layer of audio tape sticks to another.

Alignment   The relationship between the audio tape and the position of the tape-recorder head.

Alligator clips   Metal connectors used to transfer sound from a phone to a tape recorder.

Amplify   To make louder.

Amplitude   The strength or height of a sound wave or radio wave.

AMS   See *automatic music sensor.*

Analog   A recording, circuit, or piece of equipment that produces an output that varies as a continuous function of the input, resulting in degradation of the signal as material is copied from one source to another.

ATRAC   See *Adaptive Transform Acoustic Coding.*

Attack   The time it takes an initial sound to build up to full volume.

Audio card   A connection between a computer-based audio workstation and other audio equipment.

Audio chain   The route through various pieces of equipment that sound takes in order to be broadcast or recorded.

Audio console   The piece of equipment that mixes, amplifies, and routes sound.

Audio routing switcher   A type of patch panel that allows audio inputs to be switched to various outputs electronically.

Audio signal   A sound signal that has been processed into an electromagnetic form.

Audio tape recorder   A device that rearranges particles on magnetic tape in order to store sound.

Audition   An output channel of an audio console.

Automatic music sensor   A button on a digital audio tape recorder that allows the operator to skip forward or backward to the start of a new song.

Aux   See *auxiliary.*

Auxiliary   An output channel of an audio console.

Azimuth   A tape alignment problem in which the tape recorder head leans to one side or the other.

Backing layer   The back side of audio tape—the side that does not have magnetic coating.

Balance control   A knob on stereo input channels used to determine how much sound goes to the right channel and how much sound goes to the left channel.

Balanced cable   A cable with three wires—plus, minus, and ground.

Band cut filter   See *band reject filter.*

Band pass filter   A filter that cuts all frequencies outside a specified range.

Band reject filter   A filter that allows all frequencies to pass except a specified frequency range.

Bass reflex   A speaker enclosure design that has a vented port to allow rear sounds to reinforce main speaker sounds.

Bass roll-off switch   A switch that turns down bass frequencies to counter the proximity effect.

Belt-drive turntable   A system in which the turntable motor is coupled with the platter by a thin rubber belt.

Bias   A high frequency signal that improves frequency response of a recording and cuts down distortion.

Bidirectional   Picking up sound from two directions; usually refers to a microphone pickup pattern.

Binary   A number system that uses two digits, 1 and 0.

Blast filter   See *windscreen.*

Boom arm   A microphone stand for use in the radio studio, consisting of metal rods designed somewhat like a human arm; one end goes into a base that can be mounted on a counter near the audio console, and the other end supports the mic.

Boom stand   A stand that can be placed away from an announcer; usually it consists of one vertical pipe with a horizontal pipe at the top of it.

Bouncing tracks   A multitrack recording technique that combines two or more tracks and transfers them to a vacant track.

Boundary mic   See *pressure zone microphone.*

Bulk eraser   See *degausser.*

Cable   Wire that carries audio signals.

Cannon connector   See *XLR connector.*

Capacitor mic   Another name for a condenser mic.

Capstan   A metal shaft that controls the speed of a tape recorder.

Capstan wrapping   Putting masking tape around a capstan to speed up or slow down the sound recorded on or played from a reel-to-reel recorder.

Cardioid   Picking up sound in a heart-shaped pattern; usually refers to a mic pickup pattern.

Cartridge   A device that converts the vibrations from the turntable stylus into variations in voltage; also, the endless-loop tape container used in a tape recorder.

Cartridge recorder   A tape recorder that uses tape that is in an endless loop.

Cassette   A plastic case containing 1/8-inch audio tape.

Cassette recorder   A tape recorder that records and plays back 1/8-inch tape housed in a plastic case.

CD   See *compact disc.*

CD player   The piece of equipment that uses a laser to play back compact discs.

CD-R   See *CD recorder.*

CD recorder   A type of CD machine that can record as well as playback compact discs.

Channel   The route an audio signal follows; also, grouping of controls on an audio console associated with one input.

China marker   A pen-type device used to mark edit points on audio tape.

Chorusing   A multitrack overdubbing technique in which an announcer reads the same script on several different tracks to give a "chorus" effect.

Circumaural   See *closed headphones.*

Closed headphones   A ring-shaped muff that rests on the head, not the ear, through which a person can hear sound.

Closed reel   A term used to categorize audio tape recorders; refers to recorders using tape enclosed in plastic cases, such as a cassette or DAT.

Coding   In digital technology, assigning a 16-bit binary "word" to the values measured during quantizing.

Coincident miking   Using multiple microphones with pick-up patterns that overlap; usually refers to stereo miking techniques.

Cold ending   Music or song that ends with a natural, full-volume ending.

Combo   The working procedure by which the radio announcer is also the equipment operator.

Compact disc   A round, shiny disc onto which sound is recorded digitally so that it can be read by a laser.

Companders   Signal processing equipment that compresses dynamic range during recording and expands it during playback.

Compression   A sound wave characteristic that occurs when the air molecules are pushed close together; also, a system for encoding digital data bits so fewer can be placed on a disc or tape yet still represent the original data.

Compressor   A volume control usually associated with the transmitter that boosts signals that are too soft and lowers signals that are too loud.

Condenser mic   A mic that uses a capacitor, usually powered by a battery, to respond to sound.

Connector adapters   Freestanding connector parts that allow one connector form to be changed to another.

Connectors   Metal devices to attach one piece of audio equipment to another.

Control board   See *audio console.*

Copy holder   A small easel that sits on the audio console and frees the reader's hands to operate equipment.

Cross-fade   To bring up one sound and take down another in such a way that both are heard for a short period of time.

Cross-pair miking   See *X-Y miking.*

Crossover   An electronic device that sends low frequencies to the speaker woofer and high frequencies to the tweeter.

Crosstalk   The picking up on a tape track of the signal from another track.

Cue   To preview an input (such as a CD or audiotape) before it goes over the air; also, to set up an audio source at the point where it is to start.

Cue burn   Damage to the outer grooves of a record caused by backtracking the record.

Cue defeat   Switch on a cart recorder that can be set so that a cue tone is not put on the audio tape during recording; used for editing purposes.

Cue talent   A signal given to talent that means, "You're on"; it is given by pointing the index finger.

Cue tone   A tone that cannot be heard that is put on a cartridge tape to stop it automatically.

Cue wheel   Part of a CD player that allows the operator to find the exact starting point of the music.

Cupping   The turning up of the edges of audio tape.

Curling   The twisting of audio tape from front to back.

Cut   A hand signal given to talent at the end of a production; it is given by "slicing your throat" with your index finger.

DASH   See *digital audio stationary head.*

DAT   See *digital audio tape.*

DAW   See *digital audio workstation.*

dB   The abbreviation for decibel.

dbx®   A noise reduction system that compresses both loud and soft parts of a signal during recording and then expands them during playback.

DCC   See *digital compact cassette.*

Dead air   A long pause when no sound is heard.

Dead roll   To play music with the volume turned down at first to shorten the piece's duration.

Dead studio   A studio with very little echo or reverberation, caused by a great deal of absorption of the sound.

Decay   The time it takes a sound to go from full volume to sustain level.

Decibel   A measurement to indicate the loudness of sound.

De-esser   A processor that gets rid of sibilant sounds without affecting other parts of the signal.

Degausser   A magnetic unit that erases tapes.

Demagnetizer   A device to remove magnetic buildup on a tape-recorder head.

Desk stand   A mic stand for a person in a seated position.

Diffusion   Breaking up sound reflections by using irregular room surfaces.

Digital   A recording, circuit, or piece of equipment in which the output varies in discrete on-off steps in such a way that it can be reproduced without degradation of signal.

Digital audio editor   Equipment that uses standard or proprietary computers to edit sound.

Digital audio stationary head   A digital recording system that records horizontally on reel-to-reel recorders.

Digital audio tape   High-quality cassette tape that can be dubbed many times without degradation because of the sampling process of its recording method.

Digital audio workstation   A computer-based system that can create, store, edit, mix, and send out sound in a variety of ways all within one basic unit.

Digital cartridge recorder   A piece of equipment that operates similar to an analog cart machine but stores sound on a computer disc.

Digital compact cassette   A digital format that is compatible with analog cassette tapes.

Digital delay   A unit that holds a signal temporarily and then allows it to leave the unit.

Digital distribution network   A network that links ad agencies, production houses, or record companies with radio stations to deliver CD-quality audio via PC-based servers and phone lines.

Digital reverb   A unit that produces reverberation electronically.

Digital signal processor   A type of electronic audio card used for computer editing.

Digital Versatile Disc   A new video format that has the capacity to hold a feature-length movie on a compact disc-styled medium. It can also be used for music and computer data.

Direct-drive turntable   A system in which a turntable platter sits on top of the motor.

Direct sound   Sound that goes straight from a source to a microphone.

Disc jockey   A person who introduces and plays music for a radio station. The term arose because the person plays recorded discs and rides the gain on the audio board.

Distortion   A blurring of sound caused by overamplification or other inaccurate reproduction of sound.

Dolby   A noise reduction system that raises the volume of the program signal most likely to be affected by noise during production, then lowers it again during playback so that the noise seems lower in relation to the program level.

Dovetailing   A multitrack overdubbing technique in which a single announcer appears to have a dialogue with himself or herself by recording different parts of a script on different tracks.

Drop-out   A flaking off of oxide coating from audio tape so that the total signal is not recorded.

DSP   See *digital signal processor.*

DSP audio card   A necessary element in order to use standard computers to edit audio.

Dubbing   Electronically copying material from one tape to another.

Duration   The time during which a sound builds up, remains at full volume, and dies out.

DVD   See *digital versatile disc.*

Dynamic mic   A mic that consists of a diaphragm, a magnet, and coils. It is extremely rugged and has good frequency response, so it is used often in radio.

Dynamic range   The volume changes from loud to soft within a series of sounds; also, the amount of volume change a piece of equipment can handle effectively.

Dynamic speaker   A speaker with a cone attached to a voice coil. Electrical current in the voice coil creates a magnetic force that moves the cone.

Earbud   A headphone that fits in the ear.

Echo   Sound that bounces off one surface.

Editing   Splicing or dubbing material to rearrange or eliminate portions of it.

Electret microphone   A type of condenser mic with a permanently charged capacitor.

Electromagnetic speaker   Another name for a dynamic speaker.

Electrostatic headphones   Headphones that require external amplification.

Electrostatic speaker   A rarely used type of monitor speaker.

EQ   The general process of equalization.

Equalization   The adjustment of the amplification given to various frequencies like high frequencies or low frequencies.

Equalized   Audio sound that has had the amplification of its various frequencies adjusted.

Equalizer   The unit that adjusts the amount of amplification given to particular frequencies like high or low frequencies.

Equalizer/filter switch   A switch on a turntable to eliminate scratchy noises in the record.

Equal loudness principle   The hearing of midrange frequencies better than high or low frequencies.

Eraser/splice finder   A degausser that looks like a cart machine; audio carts that are inserted into it are erased and stopped just past where the tape was joined together.

Event   The start point to end point for an edit; usually associated with digital audio editors.

Fade   To gradually increase or decrease the volume of music to or from silence.

Fade-in   To bring sound up from silence to full volume.

Fade-out   To take sound from full volume to silence.

Fader   Part of an audio console that moves up and down to control volume.

Feedback   A howling noise created when the output of a sound (usually from a speaker) is returned to the input (usually a mic).

Feed reel   See *supply reel.*

Filter   A unit that cuts out a particular frequency range of the audio signal.

Flanger   A device that electronically combines an original signal with a slightly delayed one.

Flat   See *flat frequency response.*

Flat frequency response   The quality of a frequency curve wherein all frequencies are produced equally well.

Floor stand   A mic stand for a person in a standing position.

Flutter   Fast variations in sound speed.

FM microphone   A mic that does not need a cable because it consists of a small transmitter and receiver.

Four-track recorder   A machine that records four signals on a tape, all going in the same direction.

Frame   The housing for a computer chassis, power supply, and motherboard.

Frequency   The number of cycles a sound wave or radio wave completes in one second.

Frequency response   The range of highs and lows that a piece of equipment reproduces.

Full-track   A recording method that uses the whole tape for one monophonic signal.

Fundamental   A basic tone and frequency that each sound has.

Gain control   A knob or fader that makes sound louder or softer.

Gain trim   Controls on an audio board that are used to fine tune the volume of each input.

General Radiotelephone Operator License   A license given by the FCC that requires thorough knowledge of engineering and broadcast law. A test is required to obtain this license.

Give mic level   A signal given to talent to tell them to talk into the mic so the audio engineer can set controls properly. It is given by "chattering" one hand, with the palm down and the thumb under the second and third fingers.

Graphic equalizer   An equalizer that divides frequency responses into bands that can then be raised or lowered in volume.

Grease pencil   A crayonlike substance used to mark edit points on tape.

Guard bands   Small portions of blank tape between each recorded track and at the edges of the tape.

Half-track mono   The recording of two separate mono signals on a tape, one going to the left and one going to the right.

Half-track stereo   See *two-track stereo*.

Hand signals   Method of communication that radio production people use when a live mic prohibits talk or when they are in separate rooms.

Hard disk drive   A storage medium built into a computer.

Hard wiring   Connecting equipment in a fairly permanent manner, usually by soldering.

Harmonics   Exact frequency multiples of a fundamental tone.

Head   An electromagnet that rearranges iron particles on tape; also, the beginning of an audio tape.

Headphones   Tiny speakers encased in something that can be placed in, or close to, the ear.

Headshell   The front part of a turntable tone arm where the cartridge and stylus are installed.

Heads out   Having tape on a reel, with the beginning of the audio tape facing out.

Height   A tape alignment problem in which the tape head is too high or too low.

Hertz   A measurement of frequency based on cycles of sound waves per second.

Hiss   A high frequency noise problem inherent in the recording process.

Hum   A low frequency noise problem caused by leaking of the 60-cycle AC power current into the audio signal.

Hypercardioid   Picking up sound well from the front, but not the sides; usually refers to a mic pick-up pattern.

Idler arm   A tension part of a reel-to-reel tape recorder that will stop the recorder if the tape breaks.

Idler-wheel turntable   A system in which the turntable motor shaft drives a rubber disc that drives the platter.

Imaging   The apparent space between speakers and how sounds are heard within the plane of the speakers.

Impedance   The total opposition a circuit offers to the flow of alternating current.

Input selectors   Switches that are used to choose mic or line positions on an audio board.

Insert edit   See *punch in*.

Jacks   Female connectors.

Jewel box   A plastic case for a CD.

Kilohertz   A 1000 cycles per second.

Laser   An acronym for "light amplification by simulated emission of radiation"; a narrow, intense beam in a compact disc that reads encoded audio data.

Laser diode   A semiconductor with positive and negative electrons that converts an electrical input into an optical output.

Lavalier microphone   A small mic that can be attached unobtrusely to an announcer's clothing.

Leader tape   Plastic tape that does not contain iron particles to record. It is used primarily before and after the recording tape so that the tape can be threaded.

LEDE   See *live end/dead end*.

Limiter   A compressor with a large compression ratio that won't allow a signal to increase beyond a specified point.

Line level   An input that has already been preamplified.

Live bouncing   A multitrack recording technique that combines two or more tracks plus a live recording and transfers them to a vacant track.

Live end/dead end   A studio where one end of the studio absorbs sound and the other end reflects sound.

Live studio   A studio with a hard brilliant sound caused by a great deal of reverberation.

Low cut filter   A filter that eliminates all frequencies below a certain point.

Low pass filter   A filter that allows all frequencies below a certain point to go through unaffected.

M-S miking   See *mid-side miking*.

Magnetic layer   The part of the tape that contains the iron oxide coating.

Magneto-optical design   A recordable CD that records on a magnetic alloy and uses laser light to playback.

Master fader   The control that determines the volume of the signal being sent from the audio console.

MD   See *MiniDisc.*

Mic level   An input that has not been preamplified.

Microphone   A transducer that changes sound energy into electrical energy.

MIDI   See *musical instrument digital interface.*

Mid-side miking   A method of stereo miking where three mics are arranged in an upside-down T pattern.

Mini   See *miniphone connector.*

MiniDisc   A 2.5-inch computer-type disc that can hold 74 minutes of digital music.

Miniphone connector   A small connector with a sleeve and a tip.

MOD   See *magneto-optical design.*

Monaural   One channel of sound coming from one direction.

Monitor amplifier   A piece of equipment that raises the volume level of sound going to a speaker.

Monitor speaker   A piece of equipment from which sound can be heard.

Moving-coil   A type of turntable cartridge in which small wire coils are attached to the stylus and situated by a fixed magnetic structure within the cartridge.

Moving-coil mic   Another name for a dynamic mic.

Moving-magnet   A type of turntable cartridge that has a tiny magnet situated between two coils of wire wrapped around cores mounted in the cartridge.

Multidirectional microphone   A mic that has switchable internal elements that allow it to employ more than one pickup pattern.

Multiplay   A type of CD player that can hold up to 200 CDs and access material on them according to a prescribed pattern.

Multiple microphone interference   Uneven frequency response caused when microphones that are too close together are fed into the same mixer.

Multitrack recorder   A machine that can record four, eight, or more tracks, all going the same direction.

Musical instrument digital interface   A communication system that allows musical instruments and other electronic gear to interact with each other.

Mute switch   Control on an audio console that prevents the audio signal from going through a channel; similar to an on/off button.

Near-field monitoring   Placement of monitor speakers on a counter on each side of an audio console so they are extremely close to the announcer.

Noise   Unwanted sound in electronic equipment.

Noise reduction   Methods of eliminating unwanted sound from a signal.

Nondirectional   Another word for omnidirectional.

Notch EQ   An equalization technique similar to dip EQ but effecting an extremely narrow range of frequencies.

Notch filter   A filter that eliminates a narrow range of frequencies or one individual frequency.

Octave   A sound that doubles in frequency; for example, sounds at 220 hertz and 440 hertz are an octave apart.

Omnidirectional   Picking up sound from all directions; usually refers to a microphone pickup pattern.

On-air lights   A signal that comes on to indicate a live mic is on in the studio.

On-air studio   The studio from where programming is broadcast.

Open-air headphones   Devices to hear sound that fit onto the ear.

Open reel   A term used to categorize audio tape recorders; refers to a reel-to-reel recorder.

ORTF miking   A method of stereo miking where two mics are crossed in a precise manner determined by a French broadcasting organization.

Out of phase   A phenomenon that occurs when the sound wave from one mic or speaker is up and the sound wave from a second mic or speaker is down; the combined result is diminished or canceled sound.

Output selectors   Buttons that determine where a sound goes as it leaves the audio console.

Overdubbing   Adding new tracks to something that is already recorded; usually a multitrack recording technique.

Overtones   Pitches that are not exact frequency multiples of a fundamental tone.

Pan knob   The part of an audio board that controls how much sound goes to the right channel of a stereo system and how much goes to the left channel.

Pan pot   See *pan knob.*

Parametric equalizer   An equalizer that can control the center frequency and the bandwidth that will have its volume raised or lowered.

Patch bay   See *patch panel.*

Patching   Connecting equipment together through the use of jacks and plugs.

Patch panel   A board that contains jacks that can be used to make connections with plugs.

PBP   See *play-by-play.*

Peaking in the red   Modulating a signal so that it reads above 100% on the VU meter.

Pegging the meter   Operating sound so loudly that the needle of the VU meter hits the metal peg beyond the red area.

Penetration   A tape alignment problem in which the tape head is too far forward or too far back; also, sound that goes through a surface and is transmitted into the space on the other side of the surface.

Performance studio   A studio used primarily by actors or musicians that has microphones but not other production equipment.

Perspective   The spatial relationship of sound; e.g. sounds that are supposed to be distant should sound distant.

Phantom power   Power that comes from a recorder or an audio board through a mic cable to a condenser mic.

Phase   The up and down position of one sound or radio wave in relation to another.

Phone connector   A connector with a sleeve and a tip.

Phono connector   See *RCA connector.*

Photodiode   The part of a CD player that provides the data signal that will be converted to an audio signal.

Pickup pattern   The area around a mic where it "hears" best.

Pinch roller   A rubber wheel that holds tape against the capstan.

Pin connector   See *RCA connector.*

Ping-Ponging tracks   See *bouncing tracks.*

Pinning the needle   See *pegging the meter.*

Pitch   Highness or lowness of a sound determined by how fast its sound wave goes up or down.

Planar-magnetic   A rarely used type of monitor speaker.

Plastic base   The middle part of audio tape, usually made of polyester.

Plate microphone   See *pressure zone microphone.*

Plate reverb   A unit consisting of a large metal plate suspended in a frame that vibrates when a transducer changes an audio signal to mechanical energy. A mic then picks up the vibrations as reverberation.

Platter   The part of the turntable on which the record rests.

Play-by-play   A term designating sports broadcasting from the scene.

Play-level pot   A fader or potentiometer that controls the volume of a sound signal that is being played back.

Playlist   An order into which edit events are assembled; usually associated with audio tape editors.

Plugs   Male connectors.

Polar pattern   A two-dimensional drawing of a mic's pickup pattern.

Pop filters   See *windscreen.*

Pot   See *potentiometer.*

Potentiometer   A round knob that controls volume.

Preamplification   The initial stage at which volume is boosted.

Pressure mic   Another name for a dynamic mic.

Pressure pads   Small, soft elements that keep cartridge tape pressed against the tape heads.

Pressure zone microphone   A flat microphone that, when set on a table or other flat surface, uses that surface to collect the sound waves and therefore can pickup audio levels from a fairly widespread area.

Print-through   The bleeding through of the magnetic signal of one layer of tape to an adjacent layer of tape.

Prism system   The part of a CD player that directs the laser to the disc surface.

Production studio   The place where material for radio is produced before it is aired.

Program   An output channel of an audio console.

Program/audition/auxillary switch   A switch that determines where sound is sent when it leaves an audio console.

Proximity effect   A boosting of bass frequencies as a sound source gets closer to a condenser mic.

Punch in   Editing by recording over one section of a track but leaving what was before and after the edited section intact; a technique associated with multitrack recording.

PZM microphone   See *pressure zone microphone.*

Quantizing   In digital technology, determining how many levels or values each sample will be broken down into; the standard for most digital recording is 256 quantizing levels.

Quarter-inch phone   See *phone connector.*

Quarter-track stereo   The recording of two stereo signals on one tape in which two signals go to the left and two go to the right.

R-DAT   See *rotary head digital audio tape.*

Radio microphone   Another name for an FM microphone.

Rarefaction   A sound wave characteristic that occurs when the air molecules are pulled apart.

RCA connector   A connector with an outer sleeve and a center shaft.

Record-level pot   A control that adjust the volume of an incoming sound signal.

Record player   A unit that spins a record, picks up a signal, and amplifies the sound through a speaker.

Record sleeves   Heavy paper record jackets.

Reel-to-reel recorder   A tape recorder that uses open reels of tape placed on a feed reel and a take-up reel.

Reflected sound   Sound that bounces back to the original source.

Region   In digital audio editing, a common designation for a section of audio that is to be edited or saved for later use.

Regulated phase microphone   A mic that consists of a wire coil impressed into the surface of a circular diaphragm that is suspended within a magnetic structure.

Reinforced sound   Sound that causes objects to vibrate at the same frequency as the original sound.

Release   The time it takes a sound to die out from sustain level to silence.

Remote start switches   Buttons that enable a piece of equipment to be operated from a distance.

Restricted Radiotelephone Operator Permit   A license that the FCC gave to people who needed to operate equipment and keep a station on the air. It is no longer required or available.

Reverb   See *reverberation.*

Reverberation   Sound that bounces off two or more surfaces.

Reverb ring   The time it takes for a sound to go from full volume to silence.

Reverb route   The path a sound takes from a source to a reflective surface and back again.

Reverb time   See *reverb ring.*

Reverse echo   A production technique used to punctuate a key word.

RF microphone   Another name for an FM microphone.

Ribbon drive speaker   A rarely used type of monitor speaker.

Ribbon mic   A mic that consists of a metallic ribbon, a magnet and a coil. Because it is bulky, heavy, and fragile, it is rarely used in radio anymore.

Riding the gain   Adjusting volume during production.

Riding in the mud   Operating volume consistently below 20 percent on the VU meter.

Rip and read   To read news copy from the wire service machine with very little editing.

Rotary head digital audio tape   Another name for digital audio tape.

RPM   Revolutions per minute.

Sampling   In digital technology, the process of taking from the original sound source to convert to binary data.

Sampling rate   The number of times per second that a reading of the sound source is taken in order to convert it to binary data.

Scattered wind   When tape does not spool up evenly on a reel.

Sealed baffle   See *acoustic suspension.*

Sealed box   See *acoustic suspension.*

Segue   To cut from one sound at full volume to another sound at full volume.

Sel sync   Selective synchronization, a tape recorder feature that makes a record head act as a play head.

Sensitivity   A mic's efficiency in terms of volume.

Shock mount   A mic stand that isolates the mic from mechanical vibrations.

Shotgun microphone   A highly directional mic that consists of a mic capsule at one end of a tube or barrel that is aimed toward the sound source.

Shucks   See *record sleeves.*

Signal-to-noise ratio   The relationship of desired sound to inherent unwanted electronic sound. The higher the ratio, the purer the sound.

Signal processing   Manipulating elements of sound, such as frequency response and dynamic range, so that the resulting sound is different from the original sound.

Slider   See *fader.*

Slip cueing   Preparing a record to play by having the turntable motor on and holding the edge of the record until it should be played.

SMPTE time code   An electronic language developed for video that identifies each picture frame.

S/N   See *signal-to-noise ratio.*

Solo switch   A button that allows one particular audio board sound to be heard on the monitor.

Soundfile   A segment of audio recorded on a hard disc; usually associated with digital audio editors.

Soundproofing   Methods of keeping wanted sound in the studio and unwanted sound out of it.

Sound signal   A noise that has not been processed into an electromagnetic form.

Source/tape switch   A switch that allows someone to monitor either the input or the output of a tape recorder.

Spaced pair miking   See *A-B miking.*

Speaker   A transducer that converts electrical energy into sound energy.

Speaker level   An input that has been amplified several times in order to drive a speaker.

Speed selector switch   On a turntable, the control that determines whether the record plays at 33⅓ RPM or 45 RPM.

Splicing   The physical cutting of audio tape.

Splicing block   The device that holds audio tape during editing.

Splicing tape   Special tape used for holding together audio tape in the editing process.

Split-pair miking   See *A-B miking.*

Spring reverb   A coiled spring that vibrates when a transducer sends an audio signal through it. A mic then picks up the vibrations as reverberation.

Stacking   A multitrack overdubbing technique in which an announcer sings harmony to a previously recorded track.

Stand-by   A signal given to talent prior to going on-air by holding one hand above the head with the palm forward.

Standing wave   A combination of a sound wave going one direction and an identical sound wave going in the opposite direction.

Stereo   Sound recording and reproduction that uses two channels coming from right and left to imitate live sound as closely as possible.

Stereo microphone   A mic that incorporates small multiple-sound-generating elements within a single mic housing that can record sound in such a way that when it is played back, it sounds like it is coming from two areas.

Stereosonic miking   A method of stereo miking where two bidirectional mics are placed one on top of the other.

Stereo synthesizer   A device that inputs a monophonic audio signal and simulates a stereo output signal.

Stylus   A small, compliant strip of metal that vibrates in record grooves.

Supercardioid   Picking up sound well from the front but not the sides; usually refers to a mic pickup pattern.

Supply reel   The reel on the left-hand side of a reel-to-reel or cassette tape recorder that holds the tape before it is recorded or played.

Supra-aural   See *open-air headphones.*

Surface-mount microphone   See *pressure zone microphone.*

Sustain   The amount of time a sound holds its volume.

Sustain ending   Music or songs that end with the last notes held for a period of time, then gradually faded out.

Tail   The end of an audio tape.

Tails out   Having tape of a reel with the end of the audio tape facing out.

Take-up reel   The reel on the right-hand side of a reel-to-reel or cassette tape recorder that holds the tape after it is recorded or played.

Talk-back switch   A simple intercom on an audio console that allows the operator to talk with someone in another studio.

Tangency   A tape alignment problem in which the head is not pointed straight ahead.

Tape guide   A stationary pin that leads tape through the transport system of a reel- to-reel recorder.

Tape loop   A piece of audio tape spliced together so that it plays around and around in a continuous circle.

Tape transport   The part of a tape recorder that moves the tape from the supply reel to the take-up reel.

Telephone coupler   See *telephone interface.*

Telephone interface   A piece of equipment that connects telephone lines to broadcast equipment.

Tensilize   To prestretch an audio tape.

Tension arm   A moveable guide for tape on a reel-to-reel recorder.

Three-pin connector   See *XLR connector.*

Three-way system   A monitor speaker that divides sound not just to a woofer and tweeter, but also to another driver such as a midrange.

Threshold of hearing   The softest sound the human ear can hear, noted as 0 decibels.

Threshold of pain   The loudness level at which the ear begins to hurt, usually about 120 decibels.

Timbre   The distinctive quality of tone that each voice or musical instrument has.

Timer   A mechanism with a series of numbers that can be used to indicate how long something is recorded.

Tone   See *timbre.*

Tone arm   The device that holds the turntable cartridge and stylus.

Tone control A control that increases the volume of the high frequencies or the low frequencies.

Tone generator An element in an audio board or other piece or equipment that produces a tone that can be set to one hundred percent to calibrate equipment.

Tracking force The weight of a turntable tone arm.

Track sheet A format for keeping notes of what material is recorded on what tracks of a multitrack recording.

Transducer A device that converts one form of energy into another.

Tray The area where the CD sits so that it can spin and be read by the laser.

Trim See *gain trim.*

Turntable A device for spinning a record and converting its vibrations into electrical energy.

Tweeter The part of a speaker that produces high frequencies.

Two-track stereo The recording of two tracks on one tape, both going the same direction to produce stereo sound.

Two-way speaker system A speaker that has a woofer, a tweeter, and a crossover.

Type I A cassette tape that uses ferric oxide as its magnetic material.

Type II A cassette tape that uses chromium dioxide or a chrome equivalent as its magnetic material.

Type III A cassette tape that had a dual layer ferri-chrome magnetic material; it is no longer available.

Type IV A cassette tape that uses pure metal magnetic material.

Ultracardioid Picking up sound well from the front, but not the back or sides; usually refers to a mic pickup pattern.

Ultra-Harmonizer A signal processor, manufactured by Eventide, Inc., that performs a number of functions, including pitch alteration, compression/expansion, delay, reverb effects, flanging, and repeat.

Unbalanced cable Cable with two wires, of which one is positive and the other is combined negative and ground.

Unidirectional Picking up sound from one direction; usually refers to a mic pickup pattern.

Unity-gain A device whose circuits do not make the output signal louder or softer than the input signal.

User interface A device, such as a keyboard or mouse, that allows a person to interact with a computer.

Variable resistor A device that controls the amount of signal that gets through the audio console, and thereby controls the volume.

Vented baffle See *bass reflex.*

Vented box See *bass reflex.*

Vocal stripping Removing the lyrics from a song by manipulating the recorded channels.

Voice doubling A multitrack overdubbing technique in which an announcer reads the same script on two different tracks to give a double voice effect.

Voice-over Speech over something else, such as music.

Volume Loudness.

Volume control See *gain control.*

VU meter A unit that gives a visual indication of the level of volume.

Walking over Talking over the vocal portion of a song, such as when an announcer is introducing a record. Normally, an announcer only talks over the instrumental portion.

Wave envelope A representation of a total sound, including its attack, decay, sustain, and release.

Waveform The shape of an electromagnetic wave.

Wavelength The distance between two crests of a radio or sound wave.

Well See *tray.*

Windscreen A ball-shaped accessory placed over the microphone to reduce plosive sounds.

Wireless headphone A device to hear sound that transmits an RF, or infrared, audio signal from the source to the headphone.

Wireless microphone Another name for an FM microphone.

Woofer The part of the speaker that produces low frequencies.

Wow Slow variations in sound speed.

X-Y miking A method of stereo miking where two mics are placed like crossed swords.

XLR connector A connector with three prongs.

Zenith A tape alignment problem in which the head is tilted too far forward or backward.

Suggested Reading

Alten, Stanley R. *Audio In Media.* 4th ed. Belmont, CA: Wadsworth Publishing Company, 1994.

Berg, Richard E. and Stork, David G. *The Physics of Sound.* Englewood Cliffs, NJ: Prentice-Hall, 1982.

Book, Albert C., Cary, Norman D. and Tannenbaum, Stanley I. *The Radio and Television Commercial.* 2nd ed. Chicago: Crain Books, 1984.

Catsis, John R. *Sports Broadcasting.* Chicago: Nelson-Hall Publishers, 1996.

Clifford, Martin. *Microphones.* Blue Ridge Summit, PA: TAB Books, 1986.

Davidson, Homer T. *Troubleshooting and Repairing Audio Equipment.* Blue Ridge Summit, PA: TAB Books, 1987.

Day, Louis A. *Ethics in Media Communication: Cases and Controversies.* Belmont, CA: Wadsworth Publishing Company, 1991.

Hagerman, William L. *Broadcast Announcing.* Englewood Cliffs, NJ: Prentice Hall, 1993.

Hewitt, John. *Air Words: Writing for Broadcast News.* Mountain View, CA: Mayfield Publishing Company, 1988.

Hilliard, Robert L. *Radio Broadcasting: An Introduction to the Sound Medium.* White Plains, NY: Longman, 1985.

Hitchcock, John. *The Tools, Skills, and Business of Radio.* Vincennes, IN: Original Company, 1987.

Holsupple, Curtis. *Skills for Radio Broadcasters.* Blue Ridge Summit, PA: TAB Books, 1987.

Hyde, Stuart W. *Television and Radio Announcing.* 7th ed. Boston: Houghton Mifflin Company, 1995.

Johnson-Walker, Kenneth W., et. al. *The Science of Hi- Fidelity.* Dubuque, IA: Kendall-Hunt, 1986.

Keith, Michael C. *Broadcast Voice Performance.* Boston: Focal Press, 1989.

Keith, Michael C. *The Radio Station.* 4th ed. Boston: Focal Press, 1997.

Levin, Murray B. *Talk Radio and the American Dream.* Lexington, MA: Lexington Books, 1987.

MacDonald, R. H. *A Broadcast News Manual of Style.* New York: Longman, 1987.

McLeish, Robert. *The Technique of Radio Production.* Boston: Focal Press, 1988.

McSpadden, M. Rogers. *Basic Radio Production Handbook.* New York: Vantage, 1988.

O'Donnell, Lewis B., Benoit, Phillip, and Hausman, Carl. *Modern Radio Production.* 4th ed. Belmont, CA: Wadsworth Publishing Company, 1996.

————. *Radio Station Operations.* Belmont, CA: Wadsworth Publishing Company, 1989.

Oringel, Robert S. *Audio Control Handbook.* 6th ed. Boston: Focal Press, 1989.

Patterson, Philip, and Wilkins, Lee. *Media Ethics: Issues and Cases.* Dubuque, IA: Wm. C. Brown Publishers, 1991.

Walters, Roger L. *Broadcast Writing: Principles and Practice.* New York: Random House, 1988.

Watkinson, John. *The Art of Digital Audio.* Boston: Focal Press, 1992.

Wong, Michael A. *A Day in the Life of a Disc Jockey.* Mahwah, NJ: Troll Associates, 1987.

Index

Contents of Accompanying Audio CD

Chapter 1—The Production Studio

Track 1:   Demonstrate various frequencies (pitches) (1.7)

Track 2:   Play difference between simple, pure tone, and complex tone (1.7)

Track 3:   Demonstrate the difference between echo and reverb (1.8)

Track 4:   Demonstrate the difference between a live studio and a dead studio sound (1.8)

Track 5:   Demonstrate the difference between a high S/N and a low S/N (1.14)

Chapter 2—The Audio Console

Track 6:   Demonstrate what happens when a mic is put into a line input (2.4)

Track 7:   Demonstrate what happens when a CD player (line level) is put into a mic input (2.4)

Track 8:   Demonstrate an audio signal riding in the mud, peaking in the red, and pegging the meter (2.9)

Track 9:   Play digital audio recorded above 100% to hear pops distortion (2.9)

Track 10: Demonstrate bass, midrange, and treble sound (2.13)

Track 11: Demonstrate panning an audio signal—sound left, sound right, sound center, sound moving left to right (2.14)

Track 12: Play audio tone—1 kHz (2.15)

Track 13: Demonstrate a fade-in then a fade-out (2.17)

Track 14: Demonstrate a cross-fade and a segue (2.17)

Track 15: Demonstrate a cold ending then a sustain ending (2.17)

Chapter 3—Microphones

Track 16: Demonstrate omnidirectional pick-up pattern by walking around mic describing the position and talking about that pick-up pattern (3.8)

Track 17: Demonstrate bidirectional pick-up pattern by walking around mic describing the position and talking about that pick-up pattern (3.9)

Track 18: Demonstrate cardioid pick-up pattern by walking around mic describing the position and talking about that pick-up pattern (3.10)

Track 19: Demonstrate impedance mismatch—high impedance mic into low impedance input (3.12)

Track 20: Demonstrate proximity effect and then bass roll-off (3.14)

Track 21: Demonstrate feedback (3.15)

Track 22: Demonstrate difference in sound between mono and stereo signal (3.17)

Track 23: Compare the sound of b's, p's, and t's with and without a pop filter (3.19)

Chapter 4—CD Players and Turntables

Track 24: Play original analog sound then digital sound that has been dubbed about ten times (4.2)

Track 25:   Play digital sound at low then high sampling rate (4.2)

Track 26:   Demonstrate wow produced by a turntable (4.14)

Chapter 5—Audio Tape Recorders

Track 27:   Demonstrate sel sync—signal in sync and problem of signal out of sync (5.4)

Track 28:   Demonstrate the difference between 3 _ and 7 tape speed (5.6)

Track 29:   Demonstrate crosstalk (5.11)

Track 30:   Demonstrate lack of track compatibility (5.11)

Track 31:   Material for Project 1—audio with levels going up and down (Project 1)

Chapter 6—Audio Tape Editing

Track 32:   Demonstrate the sound created by print-through (6.2)

Track 33:   Play sound recorded on tape with a lot of dropouts (6.5)

Track 34:   Demonstrate edit problems—tape gap and overlap (6.11)

Track 35:   Material for Project 1—"Today's weather . . . " (Project 1)

Track 36:   Material for Project 2—announcer and music clips (Project 2)

Chapter 7—Monitor Speakers

Track 37:   Play sound from the woofer and tweeter with various cross-over frequencies (7.3)

Track 38:   Demonstrate human hearing frequency response range (7.5)

Track 39:   Play sound from speaker with good flat frequency response and with poor frequency response (7.5)

Track 40:   Play sound coming from a speaker in and out of phase (7.7)

Track 41:   Material for Project 1—good quality stereo music (Project 1)

Track 42:   Material for Project 2—mono music (Project 2)

Chapter 8—Cables, Connectors, and Accessories

Track 43:   Demonstrate an unbalanced wire picking up electric motor interference (8.7)

Track 44:   Demonstrate poorly erased audio cartridge (8.11)

Track 45:   Demonstrate announcer reading using copy holder vs. Reading "down" into console (8.12)

Chapter 9—Signal-Processing Equipment

Track 46:   Demonstrate difference between wet and dry audio signal (9.1)

Track 47:   Demonstrate EQ use to lessen hiss or hum (9.5)

Track 48:   Demonstrate filter effects—low cut/pass, band cut/pass (9.6)

Track 49:   Play sound with and without Dolby (9.8)

Track 50:   Demonstrate reverb effects vs. delay effects (9.10 & 9.11)

Track 51: Demonstrate dynamic range (9.12)

Track 52: Demonstrate effects sound through a compressor (9.13)

Chapter 10—The Digital Production Studio

Track 53: Demonstrate how tightly something can be edited with digital editing (10.

Track 54: Demonstrate how digital editing can shave off little bits to shorten something (10.8)

Track 55: Material for Project 2—short piece of music from which 5 seconds can be excerpted (Project 2)

Chapter 11—Multitrack Production Techniques

Track 56: Demonstrate overdubbing—play single track and keep adding tracks and production elements (11.3)

Track 57: Demonstrate bouncing tracks (11.5)

Track 58: Demonstrate chorusing (11.9)

Track 59: Demonstrate dovetailing (11.10)

Track 60: Demonstrate slapback echo (11.11)

Chapter 12—Production Tips, Tricks, and Techniques

Track 61: Demonstrate a tape loop (12.3)

Track 62: Demonstrate backward playback (12.5)

Track 63: Play off-speed effect of capstan wrapping (12.7)

Track 64: Demonstrate use of touch tones in production spot (12.8)